Molecules in the Early Universe acted as natural temperature regulators, keeping the primordial gas cool and, in turn, allowing galaxies and stars to be born. Even now, a simple chemistry continues to control a wide variety of the exotic objects that populate our cosmos. What are the tools of the trade for the cosmic chemist? And what can they teach us about the Universe in which we live? These are the questions answered in this engaging and informative guide to *The Chemically Controlled Cosmos*.

In clear, non-technical terms, and without formal mathematics, we learn of the behaviours of molecules in a host of astronomical situations. We study the secretive formation of stars deep within interstellar clouds, the origin of our own Solar System, the cataclysmic deaths of many massive stars that explode as supernovae, and the hearts of active galactic nuclei, the most powerful objects in the Universe. We are given an accessible introduction to a wealth of astrophysics and a comprehension of how cosmic chemistry allows the investigation of many of the most exciting questions concerning astronomy today.

T0225162

The chemically controlled cosmos

The chemically controlled cosmos

—

Astronomical molecules from the Big Bang to exploding stars

T. W. HARTQUIST

Max-Planck-Institut für extraterrestrische Physik, Garching

and

D. A. WILLIAMS

University College London

Illustrations by Richard Williams

CAMBRIDGE
UNIVERSITY PRESS

CAMBRIDGE UNIVERSITY PRESS
Cambridge, New York, Melbourne, Madrid, Cape Town, Singapore, São Paulo

Cambridge University Press
The Edinburgh Building, Cambridge CB2 8RU, UK

Published in the United States of America by Cambridge University Press, New York

www.cambridge.org
Information on this title: www.cambridge.org/9780521419833

© Cambridge University Press 1995

This publication is in copyright. Subject to statutory exception
and to the provisions of relevant collective licensing agreements,
no reproduction of any part may take place without the written
permission of Cambridge University Press.

First published 1995
This digitally printed version 2008

A catalogue record for this publication is available from the British Library

Library of Congress Cataloguing in Publication data

Hartquist, T. W. (Thomas Wilbur), 1954–
The chemically controlled cosmos: astronomical molecules from the
big bang to exploding stars / T. W. Hartquist and D. A. Williams.
p. cm.
Includes index.
ISBN 0-521-41983-2
1. Cosmochemistry.   2. Molecules.   I. Williams, D. A.
II. Title.
QB450.W55   1995
523′.02–dc20   94-25123   CIP

ISBN 978-0-521-41983-3 hardback
ISBN 978-0-521-05637-3 paperback

Contents

Contents

Preface

———

> Still, the mind of man is reluctant to consider itself as the product of chance,
> or the passing result of destinies over which no god presides, least of all
> himself. A part of every life, even a life meriting little regard, is spent in
> searching out the reasons for its existence, its starting point, and its
> source. My own failure to discover those things has sometimes inclined
> me towards magical explanations, and has led me to seek in the frenzies
> of the occult for what common sense has not taught me. When all the
> involved calculations prove false, and the philosophers themselves have
> nothing more to tell us it is excusable to turn to the random twitter of
> birds, or toward the distant mechanism of the stars.
>
> M Yourcenar in *Memoirs of Hadrian*

All visible matter in the Universe has cooled to temperatures well below those
at the Earth's surface at least once since the Big Bang. Just as the terrestrial
atmosphere, at a temperature of about 300 degrees above absolute zero,[†] is
almost entirely molecular, many of the astrophysical objects that have temperatures less than several thousands of kelvins contain large abundances of
molecules. In fact, as we show in this volume, molecules have influenced the
births and distributions of all stars and galaxies, often by serving as coolants
but in other ways as well. Some of these 'normal' astronomical objects were
the progenitors of or provided environments for the formation of more
'exotic' objects, including black holes. Molecules have, therefore, affected the
birth rates and distributions of many kinds of entities, and on the large scale
the cosmos is chemically controlled.

Chemistry also plays a significant role in the evolution of individual
astrophysical sources. For instance, temperatures in the envelopes of many
old stars drop to several thousand kelvins inducing molecule formation
which triggers the production of dust grains; these grains transmit the

---

[†] Throughout this book temperatures will be measured on a scale on which the temperature is zero
when all matter is in its lowest possible energy state. A change of 1 degree on this scale
corresponds to 1 degree Celsius, but the zero point of this scale corresponds to about −273
degrees Celsius (sometimes called absolute zero of temperature). Degrees on this scale are here
called 'kelvins'.

pressure of the stellar light which they absorb to the gaseous envelopes, powering strong winds which remove stellar mass, so that a vital active star is converted into a feeble dwarf. Many of the astrophysical sources which together comprise much of the contents of the cosmos are also chemically controlled.

The existence of molecules in these astronomical objects provides astrophysicists with considerable information about them. Each molecule absorbs and emits radiation at wavelengths that are characteristic of its species, but differences in physical conditions, including the temperature and the strength of the local background radiation field, alter the relative prominence of the different spectral features formed by such species. Thus, the response of the molecules to the physical conditions affects the observed radiations from many astronomical sources in ways that permit the diagnosis of conditions in those objects. Because the relative abundances of different chemical species are also sensitive to the local physical conditions (which as we have stated above are often controlled by the chemistry) observational determinations of the relative abundances together with theoretical understanding of the most important chemical reaction networks also probe the physical natures of the objects.

Hence, the chemistry controls the evolution of astronomical objects, *and* is a diagnostic of conditions in them. Furthermore, it is often interesting in its own right. Astrochemistry produces species that sometimes have never been manufactured in detectable quantities in terrestrial laboratories, and such species are recognized as components of astronomical gases because the detected spectra are compared to the results of theoretical studies of molecular structure and radiative processes. Many scientists are interested in the mechanisms by which these unfamiliar compounds come into existence simply because those mechanisms are fundamental and challenging to understand. The field of molecular astrophysics informs and is informed by the theoretical and laboratory work of a large number of molecular physicists and chemists. Many of the earliest investigations in molecular astrophysics were carried out by scientists whose first goal was to understand how astronomical molecules come to be so abundant. Only later were their control of their physical environments and their diagnostic utility more fully realized.

Our primary aim in writing this volume has been to provide those who are not professional astronomers with a concise introduction to how chemistry controls the properties and evolution of the astronomical environments in which it takes place. This book concerns the roles of microscopic processes in determining the remarkable variety of large scale structures and activity in the cosmos. Another goal has been to show how molecular emissions are used to study the objects in which they originate.

The first chapter of the book gives a brief account of the history of the field of molecular astrophysics, of its growth in the 1960s and 1970s and of its maturation in the 1980s, as well as of instrumental developments which will occur in the 1990s and maintain the subject's vitality into the next millennium.

Before approaching molecular astrophysics, we must become familiar with some simple basic ideas of molecular science and of astronomy. We have attempted to make this book self-contained and therefore we have included a chapter (2) covering the timescales, distances, densities, and temperatures associated with different types of astronomical objects, and also a chapter (3) dealing with basic molecular structure and chemical processes. Readers who are already familiar with these concepts may wish to skip Chapters 2 and 3. To those who must read Chapters 2 and 3 to acquire a few 'technical tools' to use while studying the more exciting material of later chapters, we offer our encouragement. We hope that they will find Chapters 2 and 3 straightforward and of rewarding utility later.

Chapters 4–11 constitute the kernel of the book. The treatment of molecules in astrophysical sources is arranged, with one exception, in roughly the sequence in which the different types of sources came into existence. Hence, we have started with the molecules present even before galaxies formed and have considered the ways in which they affected the births of galaxies and of the globular clusters. Clouds of gas form in the galaxies, stars are born in those clouds, and planetary systems arise as some of the stars form. Once a star has begun to shine it loses mass in outflows which tend to be strongest when a star is young and again when it has exhausted most of its primary nuclear fuel. Some of the material left behind after star formation and some stellar outflows contain masers, which are longer wavelength relatives of lasers, and the most violent of stellar outflows are supernovae. Chapters 4–10 tell of how chemistry has controlled the properties and dynamics of astronomical sources from the pregalactic medium to the most energetic stellar events, and how molecular radiation is important in their diagnosis.

The single break from the roughly evolutionary sequential ordering occurs near the end of the book where Chapter 11 concerns active galaxies (which include quasars, Seyferts, and starburst galaxies). Such galaxies might be treated immediately after galaxy and globular cluster formation have been considered. One ground for the break in the 'evolutionary' ordering is that it seems natural to try to understand how chemistry affects interstellar clouds in our own Galaxy before attempting the diagnosis of the active galaxies through studies of their molecular emissions. Also, in contrast to Chapters 4–9 in which many of chemistry's controlling roles are highlighted, Chapter 11 concerns exclusively the passive, though interesting, diagnostic function of chemistry. Chapter 10 is the only other chapter that is so heavily balanced towards chemistry's passive properties, and the juxtaposition of the two chapters seemed to us to be natural.

Though we have treated some problems of Solar System chemistry, we have mostly restricted ourselves to questions relevant to the formation and evolution of the Solar System when a gaseous, dusty disk still extended from near the Sun to beyond the most distant planet. We have not mentioned planetary atmospheres or surfaces except obliquely when we addressed the questions of how chemistry in the proto-Solar Nebula affected the Earth's water content and whether chemistry in protoplanets resulted in comets having the ice contents that measurements show them to possess. Further exposition on planetary atmospheres has been excluded because while attempting to present molecular astrophysics as a coherent field we wished for novelty and timeliness and because of a qualitative difference between the chemistries of most of the astrophysical environments that we have treated and those of planetary atmospheres. We have tried, as far as possible in the production of a self-contained book, to write about subjects that have never before been treated extensively at a popular level; the structures and evolutions of planetary atmospheres are subjects of chapters in many texts for nonscientists. We have tried to restrict ourselves primarily to situations in which most chemical reactions of importance arise from the collision of two bodies (each of which may be an atom, a molecule, an ion, or an electron); in contrast many interesting processes in planetary atmospheres (e.g. the reactions which remove ionized species at most terrestrial altitudes and control the Earth's atmospheric electrical properties) are three-body reactions (i.e. they involve the simultaneous collision of three rather than two particles), a consequence of the relatively high densities of the atmospheres.

In the final chapter we have summarized the reasons why molecular astrophysics is about to enter a very exciting phase. Even though we have largely focused on what is known of astronomical objects and of the chemistries occurring in them, rather than on methods of research, we hope that readers will appreciate the commonality of the basic principles on which the models of the chemistries in the different environments are based. In practice one tries to explain a few observational data for a previously unobserved source with a limited model based on the realization that the data are somehow similar to those from some more familiar source. The model becomes more complicated as the passage of time leads to the collection of more data and more analysis. Then yet another type of source is observed and the similarities between data for it and the older data induce one to use the now well-articulated model in an attempt to understand the newest data. If readers succeed in seeing the commonality they will understand a great deal about the process of modern astrophysical research.

January 1995

T W Hartquist[1]
D A Williams[2]

[1] Max-Planck-Institut für extraterrestrische Physik, Garching, Germany.
[2] Department of Physics and Astronomy, University College London, UK.

Acknowledgements

———

Excerpt from *Memoirs of Hadrian* by Marguerite Yourcenar and translated by Grace Frick. Copyright © 1954, 1963 and renewed © 1982 by Marguerite Yourcenar. Reprinted by permission of Farrar, Straus & Giroux, Inc, Martin Secker & Warburg Ltd, and Editions Gallimard.

Excerpt from *Other Inquisitions, 1937–1952*, by Jorge Luis Borges, translated by Ruth L. C. Simms. Copyright © 1964. By permission of the publisher, the University of Texas Press; additional publishing rights by permission of Penguin Books USA Inc.

Excerpt from *Letters from the Earth* by Mark Twain, edited by Bernard De Voto. Copyright © 1938, 1944, 1946, 1959, 1962 by The Mark Twain Company. Copyright renewed. Used by permission of Harper Collins Publisher Inc.

Excerpt from *The Mythic Image* by Joseph Campbell. Copyright © 1974 by Princeton University Press. Reprinted by permission of Princeton University Press.

Excerpt from *Juno and the Paycock* by Sean O'Casey. Reprinted by permission of Pan Macmillan Ltd.

1

A brief history

—

I have suspected that history, real history, is more modest and that its essential dates may be, for a long time, secret.

Jorge Luis Borges in 'The Modesty of History' in *Other Inquisitions, 1937– 1952*

The field of molecular astrophysics has grown to be a vital one through the efforts of a large number of dedicated researchers, many of whom had distinguished careers in other disciplines before pursuing the topics treated in this volume. A full history of the development of molecular astrophysics would require another book with thousands of references to works published over nearly seven decades. Here we present a very broad sketch of the evolution of the subject and emphasize the rapid observational and theoretical progress that has occurred in the last twenty years and will continue, in part due to the availability of new observing technologies, well into the next century.

1.1 The dawn of the field

In 1926 the 44 year old Sir Arthur S Eddington was invited to deliver the Royal Society's Bakerian Lecture. He was already world famous for having led the first expedition to test Einstein's general theory of relativity by measuring, during a solar eclipse, the apparent shifts (induced by the Sun's gravity) of stars with lines of sight near the eclipsed Sun's rim. In addition, Eddington had made and was to continue for two more decades to make penetrating contributions on which much of modern astrophysics is founded.

The title of Eddington's Bakerian Lecture was 'Diffuse Matter in Space'. This matter consists mostly of the interstellar gas which comprises roughly 1 per cent of the Galaxy's total mass. An additional component of the diffuse matter was recognized to be dust which in many directions obscures all but the nearest stars. The dust was known by Eddington and his contemporaries to be clumped into clouds in which most of the interstellar gas is concentrated. The masses of these clouds are now known to be up to a million times that of the Sun, so that the largest clouds are the most massive objects in the Galaxy. They are also the sites of current stellar birth.

One question addressed by Eddington in his 1926 lecture concerned whether any of these clouds, many of which are very opaque to optical light and which are typically 10^{16} times[†] less dense than the Earth's atmosphere, could contain molecules. Eddington concluded that no processes were known that could form molecules in these clouds at sufficient rates to keep their abundances high.

Yet somewhat more than a decade later, in 1937, T Dunham in the *Publications of the Astronomical Society of the Pacific*, and P Swings and L Rosenfeld in the *Astrophysical Journal*, reported their discoveries of absorption lines attributed to intervening molecules in the optical spectra of bright stars. The molecules were identified as CH, CN, and CH^+. Perhaps the most surprising aspect of the discoveries was that these molecules were detected in clouds that filter out only a moderate fraction of the optical starlight that impinges on them. This

[†] Here, 10^n means a 1 followed by n zeros. Thus, $10^2 = 100$, $10^3 = 1000$, etc. Hence, $10^{16} = 10 \times 10^6 \times 10^9$, or 10 million billion.

contrasts with Eddington's conclusion that even the darkest of the interstellar clouds would contain no molecules.

Attempts were made in the 1940s and 1950s to understand theoretically the origins of the observed molecules, but only in the 1960s were the processes that determine the abundance of molecular hydrogen, H_2, the most prevalent astronomical molecule, identified and studied quantitatively. The theoretical framework in which explanations of the abundances of CH, CN, and CH^+ and many other astronomical molecules are founded was developed in the 1970s, nearly a half century after Eddington's Bakerian Lecture was delivered. Even now (1995), the origin of CH^+ remains controversial. The field of molecular astrophysics has had its successes, as described in this volume, but many interesting problems continue to challenge researchers.

Table 1.1. *Molecules detected in interstellar and circumstellar regions.*

In fact, since many of these molecules have also been detected with the major isotope (e.g. the most abundant form of carbon atom has a mass of 12 atomic mass units and is denoted ^{12}C) substituted by a minor isotope (e.g. ^{13}C), the actual number of detected molecular species is much larger than shown in the list below. For example, CO has been detected in interstellar clouds with all possible combinations of ^{12}C and ^{13}C, and ^{16}O (the main isotope of oxygen) and ^{17}O and ^{18}O (the minor isotopes of oxygen). As another example, in cyanoacetylene, HC_3N, a linear molecule, the ^{13}C isotope can be found in any of the three positions for carbon. The hydrogen atoms in these molecules may also be substituted by deuterium (or heavy hydrogen, ^2H or D). The 'l' and 'c' forms of certain molecules, e.g. C_3H, refer to linear and cyclic geometries.

No. of atoms						
2	3	4	5	6	7	≥ 8
H_2	H_2O	NH_3	HC_3N	CH_3OH	HC_5N	CH_3OCHO
OH	H_2S	H_3O^+	C_4H	CH_3CN	CH_3CCH	CH_3C_3N
SO	SO_2	H_2CO	CH_2NH	CH_3SH	CH_3NH_2	HC_7N
SiO	NH_2	HNCO	CH_2CO	NH_2CHO	CH_3CHO	CH_3OCH_3
SiS	N_2H^+	H_2CS	NH_2CN	CH_3NC	CH_2CHCN	CH_3CH_2OH
NO	HNO	HNCS	HOCHO	HC_2CHO	C_6H	CH_3CH_2CN
NS	HCN	C_3N	$c\text{-}C_3H_2$	H_2C_4		CH_3C_4H
HCl	HNC	$c\text{-}C_3H$	CH_2CN	C_2H_4		HC_9N
PN	C_2H	$l\text{-}C_3H$	H_2C_3	$H_2C_3N^+$		$HC_{11}N$
NH	HCO	C_3S	CH_4			CH_3C_5N
CH^+	HCO^+	C_3O	HC_2NC			$(CH_3)_2CO$
CH	OCS	C_2H_2	SiH_4			
CN	HCS^+	$HOCO^+$				
CO	C_2S	$HCNH^+$				
CS	C_2O					
C_2	NaCN					
CO^+	$c\text{-}SiC_2$					
SO^+	MgNC					
SiN						
NaCl						
KCl						
AlF						

Fig. 1.1. The Orion Nebula (copyright 1981 AAT Board).

1.2   New windows on the molecular composition of the cosmos

The 1960s and 1970s were an era of great theoretical development in molecular astrophysics. However, no scientific area expands in the absence of data; one of

Fig. 1.2. The James Clerk Maxwell Telescope. The JCMT, which is located on a mountain in Hawaii and operated by the UK's Particle Physics and Astronomy Research Council on behalf of a British–Canadian–Dutch consortium, is one of several submillimetre telescopes which permit investigations of molecular emission in a previously unaccessed wavelength range. (Photograph courtesy of Dr A I Harris.)

the major reasons for the theoretical advances of the 1960s and 1970s was the concurrent (and often somewhat earlier) appearances of new methods for observing astronomical sources.

A huge boost to molecular astrophysics occurred with the first radio wavelength detection of an astronomical molecule. In 1963 a detection was made of several spectral features, at wavelengths of about 18 cm, due to absorption by foreground OH molecules of radiation emitted by a remnant caused by the interaction of an exploded star (a supernova) and the interstellar gas around it. As well as absorbing radiation, OH molecules were also observed in emission. In some places OH molecules were found to radiate with roughly ten thousand million (ten billion) times the intensity that was expected. The tremendous strengths of these OH emissions were recognized in the mid-1960s to be due to the action of processes very similar to those operating in lasers, then only recently invented in the laboratory.

Many molecules have rich spectra at wavelengths of a fraction of a millimetre up to tens of centimetres. Radio astronomy's original contributions came from studies at wavelengths of several centimetres to tens of centimetres, but during the 1970s millimetre wave astronomy began to flourish as receivers and dishes designed to work at shorter wavelengths became available. Many tens of molecular species were detected in astronomical sources for the first time in the 1960s and 1970s. The list of astronomical molecules observed at radio centimetre and millimetre wavelengths includes simple molecules like CO (carbon monoxide) as well as complicated ones such as CH_3SH (methyl mercaptan) and $HC_{11}N$ (cyano-decapenta-yne). A list of molecular species that have been detected in interstellar and circumstellar regions is given in Table 1.1.

H_2, in part because of its lightness compared to other molecules and because of its symmetry, does not have a rich spectrum at centimetre and millimetre wavelengths. Many of its lowest energy transitions lie in a range of infrared wavelengths[†] at which the Earth's atmosphere is opaque, while its highest energy transitions lie at ultraviolet wavelengths[‡] at which atmospheric absorption is also a problem. Since hydrogen is the most abundant element, being about ten and a thousand times more prevalent than helium and oxygen (the next most abundant elements), respectively, astronomers anticipated that far more H_2 than any other molecule would exist in most astronomical sources. · Even before it was observed, a detailed, quantitative theory for the production and destruction of astronomical H_2 was developed in the 1960s. Then in 1970 H_2 outside the Solar System was detected for the first time. An ultraviolet spectrograph was launched above the atmosphere on a rocket, and absorption due to interstellar H_2 of ultraviolet emission from nearby bright stars was measured. In 1973 the highly successful operation of the *Copernicus* satellite which carried an ultraviolet spectrometer made possible revolutionary studies of interstellar molecular clouds, as well as nonmolecular sources including the

---

[†] The infrared region of the spectrum is generally considered to extend from wavelengths of about 1 micrometre (or one millionth of a metre) to hundreds of micrometres. The features of longest wavelength in the spectrum of molecular hydrogen occur around 20 micrometres.

[‡] The ultraviolet region of the spectrum extends from wavelengths of about 0.3 micrometres to 0.01 micrometres. The shortest wavelengths associated with molecular hydrogen are around 0.1 micrometres which may also be written 100 nanometres (i.e. 100 nm, the prefix 'nano' meaning division by 10^9).

hot interstellar gas (with a temperature of about one million kelvins) and the atmospheres and winds of the brightest ultraviolet emitting stars.

In 1976 infrared emission from astronomical H_2 was observed for the first time. Special infrared detectors were used on a ground based optical telescope to search at wavelengths around 2 micrometres, one of the narrow infrared bands at which the Earth's atmosphere is not too opaque. Radiation was seen from moderately excited states of H_2 populated in the warm gas (with a temperature of several thousands of kelvins) in the vicinity of young stars in the Orion Molecular Cloud near the Orion Nebula which is pictured in Fig. 1.1.

In the next few years, spectroscopic satellites will permit the study of infrared emission and absorption by many other molecules. In the early 1990s, the use of precisely formed large metallic dishes with diameters of 10–15 m and of special spectroscopic receivers have made possible the study of molecular spectral features at wavelengths of several tenths of a millimetre. The James Clerk Maxwell Telescope, one of those new submillimetre instruments, is shown in Fig. 1.2. Its surface is thinly plated in aluminium, a necessary part of the design to allow the maintenance of sufficient smoothness to work at these short wavelengths. Also, arrays of dishes that operate together to map at high angular resolution sources of molecular emissions in the millimetre wavelength range have become available, and are steadily collecting data for star forming regions and other types of sources. Exciting new discoveries are anticipated from this developing technology.

Selected references

Dalgarno, A: 'The Interstellar Molecules CH and CH^+', in *Atomic Processes and Applications*, eds. Burke, P G and Moiseiwitsch, B L, North-Holland Publishing Company, Amsterdam (1976).

Duley, W W and Williams, D A: *Interstellar Chemistry*, Academic Press, London (1984).

Field, G B, Somerville, W B, and Dressler, K: 'Hydrogen Molecules in Astronomy', *Annual Reviews of Astronomy and Astrophysics*, vol 4, p. 207 (1966).

Hartquist, T W (ed.): *Molecular Astrophysics – A Volume Honouring Alexander Dalgarno*, Cambridge University Press, Cambridge (1990).

2

Setting the astronomical scene

——

Man has been here 32 000 years. That it took a hundred million years[†] to prepare the world for him is proof that is what it was done for. I suppose it is. I dunno. If the Eiffel Tower were now representing the world's age, the skin of paint on the pinnacle knob at its summit would represents man's share of that age; and anybody would perceive that the skin was what the tower was built for. I reckon they would, I dunno.

Mark Twain in 'Was the World Made for Man' in 'The Damned Human Race' in the collection *Letters from the Earth*

Our most immediate impression of the world around us is one of stability. In our literature, the mountains symbolize strength and permanence; the seas eternity. The Sun and stars seem immutable, fixed since Creation. Of course, we know that this apparent constancy is illusory. The Universe is continually evolving. The human attention span is so brief that nothing changes significantly in the lifetime of an individual, except for minor local changes, and – occasionally – the explosion of an evolved star in a supernova.

But the evidence of variation can always be found. Even on the Earth, the signs of changes in the atmosphere induced by a mere century of technological activity are there to be read. We know that the changes themselves will not be significant for some decades to come. But the evidence is there, although not immediately apparent. So it is with astronomy. On a casual inspection the Universe may appear unchanging, but the manifestations of development, i.e. of evolution, are there to be interpreted. In this book we are particularly concerned with the evidence provided by chemistry. We shall see that radiation from the molecular products of chemistry generally allows us to probe the cooler and denser regions of the Universe, which may be optically inaccessible. The study of molecules, usually through spectral lines in the infrared and radio regions, therefore complements traditional optical investigations and gives a new perspective.

[†] Current estimates are around 4500 million years, but the ratio of the estimates of the age of the human species and the Earth's age have remained approximately constant since Twain wrote.

Table 2.1. *Ages and lifetimes, in years and in 'A-units' where 10^6 y is 1 'A-day'.*

Age of the Universe	15×10^9 y	41 A-years
Age of the Galaxy	12×10^9 y	33 A-years
Age of the Sun	5×10^9 y	14 A-years
Age of the Earth	4.5×10^9 y	12 A-years
Rotation period of the Galaxy	10^8 y	3 A-months
Life of a molecular cloud	10^8 y	3 A-months
Life of a bright star	10^6 y	1 A-day
Duration of a cool envelope	10^4 y	14 A-minutes
Duration of a planetary nebula	10^4 y	14 A-minutes
Duration of human civilization	5×10^3 y	7 A-minutes
Duration of human technological era	10^2 y	9 A-seconds
Duration of a supernova	1 y	0.1 A-seconds

We shall also discover that chemistry not only allows us to trace the presence of matter in the Universe, but that it actually *controls* the evolution on all scales throughout the Universe. Were it not for the consequences of chemistry occurring at crucial times in the development of the Universe, it would certainly not be as we find it today.

2.1 Timescales

There are two kinds of timescale with which we must be concerned. One is the timescale of events in astronomy; the other is the timescale of events in the chemistry of astronomy. Both these timescales are quite different from our terrestrial experience.

The age of the Universe, i.e. the time elapsed since the Big Bang, is some 15 billion years (we use 1 billion $= 10^9$). The Sun is 5 billion years old, and the Earth is slightly younger at 4.5 billion years. The Galaxy, of which the Sun and its family of planets are members, rotates once every 100 million years. The Sun is evidently long lived. More massive stars burn much more brightly, and though they have more fuel, the largest can only last about a million years.

'Only a million years' – of course, that is an enormous, incomprehensible period of time by human standards. It is the length of time that humanoid beings have been present on Earth; however, human civilizations have been active on this planet only for the last few thousand years. A million years is, nevertheless, a useful period of time in considering astronomical and geological terms. Many processes within the Galaxy occur on this kind of timescale. Let us, therefore, regard a million years as a period of one '*astronomical day*', an '*A-day*', and relate ages and periods of some astronomical events to this new scale of time. Table 2.1 gives the times in years and in A-units associated with some astronomical and historical events. This helps to put timescales into a reasonable perspective.

The table suggests that the Universe is 41 A-years old, while – at the other extreme – human activities have been so recent as to be almost insignificant.

In astronomical terms, bright stars come and go like flowers that bloom for an A-day or two, while each of the 'plants' that produce them – the molecular clouds – remains in being for a season of several A-months. Low-mass stars, like the Sun, are much more common and much more durable. Like the grass in the garden lawn, they can survive for A-decades. The spectacular events that occur in the lives of some stars: a supernova explosion, the development of extensive cool envelopes, the blossoming of a planetary nebula, are really very brief phenomena indeed. Like dandelion seeds carried off by a puff of wind into the garden, these events are soon over.

The chemical timescales are also very important. For two species, atoms or molecules, A and B, to react together and make products requires – as a minimum condition – that the species come into reasonably close contact. If we say that 0.3 nanometres is reasonably close and that most species move with speeds around a few hundred metres per second then the time for a particular species A to be in collision with any of species B is about $10^{16}/n(B)$ seconds, where n(B) is the number of species B per cubic metre. In the Earth's atmosphere, the abundant species such as molecular nitrogen and molecular oxygen have number densities on the order of 10^{25} per cubic metre (m^{-3}), so the time between collisions of these species is really very short, about 10^{-9} seconds, a billionth of a second. However, in the Universe at large the terrestrial conditions are so unusual as to be freakish. We shall see in later chapters (see also Section 2.3) that number densities of the most abundant species, hydrogen, may range from, say, $10^3 \, m^{-3}$ in protogalaxies to $10^{20} \, m^{-3}$ in the disks of protostellar nebulae. The range of collisional timescales is, consequently, also very wide. At the lowest densities, collisions of one hydrogen atom with another occur on the average at intervals of 10^{13} seconds, nearly a million years. (A useful approximate conversion is 1 year equals 30 million seconds.) Collisions with other elements are much rarer than this.

Only a fraction of collisions usually leads to reaction of the partners and formation of the products, so at low densities nothing of chemical importance can happen until very long times have elapsed. In stellar atmospheres, however, each molecule may be subjected to hundreds of collisions per second, and chemistry can be rapid.

However, much of the chemistry in astronomy occurs slowly, and local physical conditions (e.g. temperature, density) may change faster so that the chemistry can never 'catch up'. For example, the conversion of atomic hydrogen to molecular hydrogen

$$2H \rightarrow H_2$$

is generally rather slow. A cloud of gas in which much of the hydrogen is still atomic may collapse under its own weight so quickly that even when its density has increased to a value at which one would expect much of the hydrogen to be molecular, the chemistry may not keep pace, and the content of atomic hydrogen may be greater than expected. On the other hand, if the chemistry is very fast, as in stellar atmospheres, then even if the local physical conditions

Table 2.2. *Distances in the Universe.*

Mean Earth radius	6.371×10^6 m	0.02 light seconds
Mean Earth–Moon distance	3.844×10^8 m	1.3 light seconds
Mean solar radius	6.960×10^8 m	2.3 light seconds
Mean Earth–Sun distance	1.495×10^{11} m	8.3 light minutes
Mean Jupiter–Sun distance	7.77×10^{11} m	43 light minutes
Mean Pluto–Sun distance	5.91×10^{12} m	5.5 light hours
Sun–α Centauri distance		4.3 light years
Sun–Orion Nebula distance		1.4 thousand light years
Typical separation between bright star groups		~3 thousand light years
Sun–Galactic Centre distance		27 thousand light years
Galaxy radius		50 thousand light years
Galaxy–Large Magellanic Cloud distance		170 thousand light years
Typical galaxy–galaxy distance in the local cluster		500 thousand light years
Galaxy–Virgo cluster distance		62 million light years
Galaxy–Hydra cluster distance		3.3 billion light years
Radius of the visible Universe		15 billion light years

change the chemistry responds rapidly enough to be in some kind of steady state with the physical conditions.

2.2 Distances

Astronomical distances are simply vast by Earthly standards. The usual measuring units soon seem inappropriate outside the immediate terrestrial environment (for which, of course, they were designed). Fortunately, Nature provides us with a convenient alternative measure: the light travel time. For example, the mean Earth–Sun distance is about 150 billion metres. Light takes about 500 seconds, or just over 8 minutes, to travel this distance, so we say the Sun is about 8 light minutes distant from Earth. When we come to consider distances to objects outside the Solar System, then the light travel time is a very practical measure. Table 2.2 gives some important astronomical distances and length scales. For example, the nearest star, α Centauri, is about 4.3 light years distant (a light year is 9.461×10^{15} m, or 9.461 million million kilometres). This way of measuring distance has the very practical meaning that the light we receive from the star α Centauri set out on its journey about 4 years ago. The distance to the Orion Nebula – a spectacular region (see Fig. 1.1) where massive new stars are being formed now – is more than 1000 light years, about 1.4×10^3 light years.

There is an enormous number of stars, about 100 billion, or 10^{11}, in our Galaxy. These stars are confined to a thin disk, which rotates about once every 100 million years. The radius of this disk is about 100 thousand light years. The Sun is not in any privileged position in the Galaxy but is in the galactic 'suburbs', about 30 thousand light years from the centre. The thickness of the disk is about 1 thousand light years.

11

The Galaxy has neighbours, the galaxies of the local cluster, of which the nearest is the galaxy called the Large Magellanic Cloud. Within the local cluster, the separations between galaxies are measured in hundreds of thousands of light years. Distances to other clusters of galaxies are very much greater than this, and are typically tens or hundreds of millions of light years. For example, the distance to the Virgo cluster of galaxies is 62 million light years. The visible Universe, i.e. the distance to the furthest objects that optical telescopes can detect, is some 15 000 million light years (or 15 billion light years).

The origin of this structure in the Universe, on all scales, is a major challenge to the human intellect. We argue in this book that at crucial times chemistry has played, and continues to play, a controlling role. It has affected galaxy formation (Chapter 4) and star formation (Chapter 6), in particular.

2.3 Temperatures and densities

Molecules do not generally survive in gas that is more than about ten times hotter than the Earth's atmosphere. Hence, most of the astronomical sources that we shall consider have temperatures equal to or less than several thousand kelvins. While most of the Sun's surface (at about 6000 kelvins) is warmer than this, sunspots and the envelopes of many highly evolved stars are marginally cool enough for molecules to exist. Much of the interstellar gas that is molecular has temperatures of only some tens of kelvins. We shall also describe interstellar sources in which shocks and other energy dissipation mechanisms maintain gas at several thousand kelvins. These interstellar molecular regions may be embedded in a hot gas which contains no molecules, but which exerts a pressure on molecular sources. This hot gas has temperatures of tenths of millions to tens of millions of kelvins; such temperatures are similar to that (around 10^7 kelvins) at the centre of the Sun. The reader may find it useful to recall that the temperature of a gas is proportional to the average kinetic energy of a free electron, ion, atom, or molecule. The average thermal speed rises proportionately to the square root of the temperature and is around $1 \, \mathrm{km \, s}^{-1}$ for a hydrogen molecule in a 100 kelvin gas.

While the temperatures that are associated with molecular gas in the Universe vary by a factor of less than a thousand (from less than 10 kelvins to more than 3000), the densities vary by many orders of magnitude. The number of molecules in a cubic metre of air under standard atmospheric conditions on Earth is about 3×10^{25}. One of the densest astronomical molecular environments that we shall discuss is that in the protosolar disk near where Jupiter formed. The number density there was about $10^{20} \, \mathrm{m}^{-3}$. The number densities in many of the extended envelopes around stars, say where H_2O masers are prevalent (see Chapter 9), are only around 10^{14}–$10^{15} \, \mathrm{m}^{-3}$. Of course, as stars are born the protostars get even denser than the Earth (e.g. at the Sun's centre, the number density is about $10^{32} \, \mathrm{m}^{-3}$), but the densest protostellar objects to be observed have number densities of H_2 in the range of 10^{10}–$10^{13} \, \mathrm{m}^{-3}$. Typical number densities of interstellar clouds are around

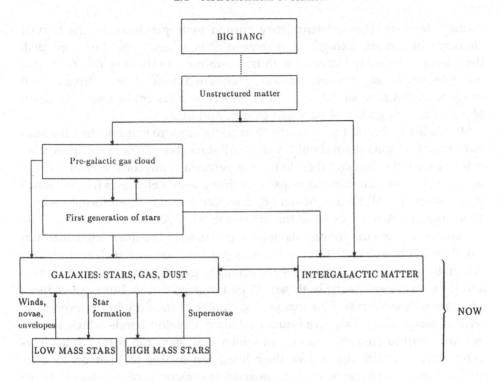

Fig. 2.1. The fate of a parcel of matter during the life of the Universe.

10^8–10^9 m^{-3}, though (as mentioned above) they are higher in these protostellar regions. The lowest density environment containing molecules that we shall consider will be the protointergalactic medium just before the first galaxies 'switched on'; the average number density of this medium then was probably smaller than 10 m^{-3}, roughly the average number density in the Universe at that time.

It is remarkable that regions with such a broad range of physical conditions can be understood. This understanding has been the product of patient observation of and diagnosis with astronomical spectra, and scientific logic tempered with inspired guesswork.

2.4 Astronomical evolution

The Universe probably began violently with a Big Bang. When it was only about three minutes old nuclear reactions, some of which formed most of the Universe's helium, prevailed. After 500 thousand years the general expansion of the Universe had caused the matter–radiation mixture filling it to cool enough that electrons interacted with positive ions to produce neutral matter, mostly atomic hydrogen and helium. The fate of this matter is illustrated in Fig. 2.1.

Some of the matter accumulated into pregalactic gas clouds through the influence of molecular hydrogen which acted as an effective cooling agent. Within these regions, the first generation of massive, short lived stars was formed. In their short existence they burned hydrogen to make the heavier elements of carbon, nitrogen, and oxygen. When these first stars·exploded, they seeded the Early Universe with trace amounts of these elements, so that the pregalactic gas became gradually enriched in carbon, nitrogen, and oxygen. In this new situation, a wider variety of stars could form: the cloud of gas became a galaxy of stars, gas clouds, and dust.

Stars with the highest mass (some 50 or 60 times more massive than the Sun) burn their fuel quickly (in about 1 A-day). All stars about 10 or more times more massive than the Sun end their lives in supernova explosions, leaving behind neutron stars and injecting more gas containing heavy elements into the interstellar medium. All the bright stars that we can see have been formed within about the last A-week or so of the life (some 30 A-years) of the Galaxy. The explosions of the supernovae maintain a pressure in the interstellar medium which helps to accumulate very tenuous gas into somewhat denser clouds. When these clouds become massive enough, gravitational collapse ensues and new stars form. Some of these will end in supernovae, but most of them are stars of lower mass. Low-mass stars, similar to the Sun, in the process of settling down into a long lived quiescent state, develop winds which stir the gas and help to create conditions in which new stars can form. Older low-mass stars, towards the end of their lives, are unable to hold on to their atmospheres which begin to drift away in enormous cool envelopes. These envelopes are ultimately lost when those objects have their moment of glory (a quarter of an A-hour) as planetary nebulae, one of which is pictured in Fig. 8.5. The remnant is generally a white dwarf star.

Our view of the Universe, then, is of a collection of galaxies set in an invisible intergalactic medium. Within our own Galaxy (and others) there are transient massive bright stars, long lived low-mass stars, and gas and dust in interstellar space. Some of the gas is hot and tenuous, some compressed into denser clumps from which new stars, both low-mass and massive, are formed. The role of chemistry in all of this is what we describe.

Selected reference

Shu, F H: *The Physical Universe: An Introduction to Astronomy*, University Science Books, Mill Valley, California (1982).

3

The tools of the trade

Chemistry has been termed by the physicist as the messy part of physics, but that is no reason why the physicists should be permitted to make a mess of chemistry when they invade it.

Frederick Soddy

How does chemistry take place outside the atmospheres of planets? How is it different from chemistry on Earth? To answer these questions we take a very crude look at the natures of atoms and molecules and how they interact with each other and with radiation.

3.1 Atomic energy levels: rules and regulations

Around 80 per cent of the mass of visible matter in the Universe is hydrogen. A hydrogen atom is the simplest type of atom and consists of a proton and an electron interacting electromagnetically. A naive model of a hydrogen atom is a picture of an electron orbiting about a proton. However, such a classical description is really inappropriate, and quantum mechanics must be used to understand the structure of an atom. Quantum mechanically, a hydrogen atom is viewed as an electron–proton system that can be in particular states in which the energy, electron angular momentum, and the sum of the spins of the proton and electron are conserved.

The first triumph for quantum mechanics was the explanation of the spectrum of the hydrogen atom. Heisenberg and Schrödinger, each using his own independently developed formulation of quantum mechanics, showed, as had already been deduced from the laboratory spectrum of hydrogen, that a hydrogen atom has many states with well-defined discrete energies as well as states that have a continuous range of energies. The theoretical work demonstrated that all states for which the average separation between the electron and proton is finite (i.e. the electron is bound to the proton) have discrete energies, and that all states for which the average proton–electron separation is infinite

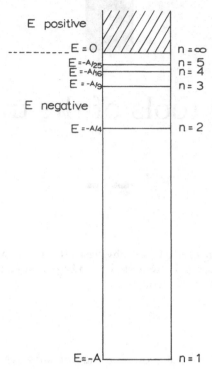

Fig. 3.1. Energy levels of the hydrogen atom.

(i.e. the electron is not bound to the proton) have energies that lie in a continuous range. That is: atomic hydrogen has *bound states with discrete energies* and *unbound states with a continuum of energies*. This discrete bound state–continuum unbound state energy structure is a general property of quantum systems (indeed of all systems – the appearance of continuity in bound classical systems is an illusion and due to the very small differences in the properties of different states compared to macroscopic properties of the classical system). Figure 3.1 shows the allowed energies of a hydrogen atom. The negative energies that are permitted in the hydrogen atom are found to take the discrete values expressed in a simple formula

$$E_n = -A/n^2,$$

where A is a positive constant energy, and n is a whole number, i.e. $n = 1, 2, 3, \ldots$. The value of E_n with $n = 1$ gives the most negative energy, $E_1 = -A$. This is the energy of the most strongly bound hydrogen atom, and that state of the atoms with $n = 1$ is called the *ground state*. If an atom in this ground state were given an amount of energy equal to or larger than A, then the total energy would be zero or positive, and the electron and proton would become infinitely separated. The value $n = 2$ gives the energy of that state to be $-A/4$, and $n = 3$ corresponds to an energy $-A/9$, and so on. The larger n becomes, the closer the negative energy E_n approaches the zero level. A hydrogen atom has, therefore, a ladder of energy levels. The atom is

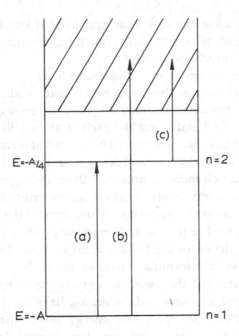

Fig. 3.2. Excitation and ionization of the hydrogen atom.

found only on one of these steps of the ladder. It cannot rest between the steps, with an energy that does not correspond with one of these steps.

However, the atom can move up and down the ladder of energy levels. If it jumps down from level $n = 2$ to level $n = 1$ then it must give out energy of an amount $E_2 - E_1$, or $\frac{3}{4}A$. A downward jump in energy takes place when accompanied by the emission of a unit of radiation called a *photon*. A collection of photons makes a beam of radiation whose frequency depends directly on the energy of the photons. For radiation emerging from a collection of hydrogen atoms, all jumping from level $n = 2$ to level $n = 1$, the frequency of the radiation is at 2.46×10^{15} hertz and the wavelength of the radiation is at 121 nanometres in the ultraviolet. The jump from $n = 3$ to $n = 2$ gives radiation with a wavelength in the visual region of the spectrum, at 656 nanometres. This is a red line, and is responsible for the red colour of many astronomical nebulae (see e.g. Fig. 1.1 which shows the Orion Nebula, where hot gas is energized by powerful nearby young stars).

Conversely, hydrogen atoms in state $n = 1$ can jump *up* to state $n = 2$ by absorbing energy from radiation with the appropriate wavelength, i.e. 121 nanometres. We say that an individual atom *absorbs* a unit of that radiation, a photon, which has exactly the right energy to allow the atom to jump from state $n = 1$ up to state $n = 2$.

Hydrogen atoms can jump not only from one bound level to another, but also from a bound level (with negative energy) to a state with positive energy so that the electron is free and not bound to the proton, as shown in Fig. 3.2. In the diagram, jump (a) requires the absorption of a photon of energy $3A/4$, while jump (b) requires the absorption of a photon of energy greater than A. Ionization can also occur from excited levels. Figure 3.2 shows ionization occurring from level $n = 2$, in jump (c). Obviously, to ionize the atom from this level takes much less energy.

In a cool gas of hydrogen atoms, most are in level $n = 1$, the ground state. However, if we heat the gas, the atoms move faster and collisions are more energetic. A few of the collisions may be energetic enough to excite one of the atoms into the $n = 2$ state, and the excited atom will then radiate as it returns to the $n = 1$ state. *So a property of a hot gas is that it radiates at wavelengths that are characteristic of the atoms (or molecules) present.*

Other atoms are much more complicated than hydrogen. Helium has two electrons so that the energy levels of helium atoms must allow for the excitation of one or both electrons, and must take into account the repulsion between the two electrons as well as their attraction to the positive nucleus. The calculation of the energy levels of the helium atom is more difficult than that of hydrogen, and no simple formula exists, as for hydrogen, to describe the energy level structure. But the essential point is the same as for hydrogen. Helium atoms, indeed all atoms and molecules, have energy levels that have discrete, i.e. distinct, values, and also energy levels that are continuous. Jumps upwards or downwards between the discrete levels give rise to characteristic energies of photons, and – therefore – characteristic wavelengths for the atoms or molecules involved. The lines are *signatures* of the atoms. For example, the line at 656 nanometres is a characteristic of hydrogen atoms and no others. *The ability to identify atoms and molecules through features in the spectrum is the basic tool of astrophysics.*

There are many examples of such lines that are often used by astrophysicists. They occur in all regions of the electromagnetic spectrum, not only in the visible. For example, another important line of atomic hydrogen occurs in the radio region and has a wavelength of 21 cm. This line has been particularly important for tracing atomic hydrogen throughout the Galaxy. Helium is the second most abundant element in the Universe, after hydrogen. Transitions to and from the ground state of helium are very energetic, and the corresponding lines fall in the far ultraviolet, near 58 nanometres. Other atoms possess characteristic optical features. For instance, a phenomenon familiar in the laboratory is the orange emission from sodium containing compounds. Common salt (sodium chloride) in a flame gives rise to an orange glow, corresponding to a wavelength near to 589 nanometres. Sodium atoms in space between us and a distant star *absorb* at the same wavelength.

3.2   Molecular energy levels: new freedoms

Let us imagine what happens when two hydrogen atoms approach each other from a great distance. Since each atom is made up of a proton and an electron, each has an electric field that is felt by the other. Each atom readjusts very slightly in this field in such a way that the atoms attract each other. If the atoms are far from one another, then on average only one of the electrons is between the

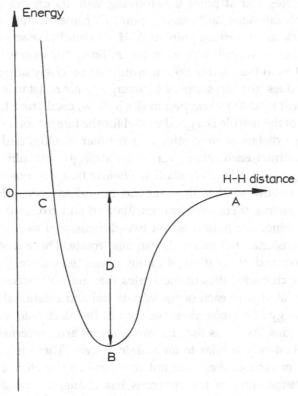

Fig. 3.3. The interaction of two hydrogen atoms.

two protons while the other electron is on the far side of one of the protons. This 'in-between' electron partly screens the protons from each other, but – on average – in the region between the two protons the electric field is slightly dominated by the protons and, hence, attractive to electrons. This attraction for the electrons slightly distorts the electron distributions enhancing the maximum of the distributions near the midplane of points equidistant from the two protons. The increase in the electron concentration between the protons further shields the protons from one another and even creates a pull, towards the midplane, on the protons. As the protons become closer the distortion of the electron distribution becomes greater, and the pull towards the midplane on the protons increases. As the atoms become closer, the electrons on each respond more to both protons. They no longer 'belong' to one of the atoms but to the pair, since they, and the protons, are identical. If the protons come even closer together there is a high probability that neither electron is between the protons; then the protons are not screened from one another and their electric interaction causes them to repel one another. The potential energy of a proton as a function of the proton–proton separation in a ground state hydrogen molecule is shown in Fig. 3.3. We can interpret this diagram in the following way.

Imagine rolling a marble into a hollow represented by the curve ABC. Initially, at point A, the marble moves slowly, but as the curve steepens the marble accelerates and at point B is moving with its greatest speed. It then begins to climb the slope and reaches point C, comes to rest, falls back to B and climbs back to its starting point at A. If we could take some energy away from the marble, it would remain in the hollow. For example, if we could take energy D from the marble then it would sit precisely at point B. In practice, a marble does give up some of its energy to air resistance and to rolling friction, and will tend to be trapped in the hollow, oscillating to and fro.

The motion of the marble is a good model for the interaction of two hydrogen atoms. At large distances they attract each other weakly, and as they come together they attract each other even more strongly; but ultimately at close range they repel each other. We shall see below that it is possible to trap the atoms in the hollow so that they remain bound to each other, forming a molecule. In general, then, they will oscillate to and fro, as does the marble in the hollow. Since the hollow is not two-dimensional as drawn in Fig. 3.3, but three-dimensional, the molecule can also rotate. These motions of atom–atom oscillation and of rotation of atoms about axes through the centre of mass are those characteristics of molecules that are not shared by atoms. An energy is associated with each of the vibrational and rotational motions.

The total energy of a molecule is the sum of the electronic, vibrational, and rotational energies. As far as the electron energies are concerned, a molecular system is qualitatively similar to an atomic system. There is a set of discrete energy levels; of course, these are not the same as the atomic energy levels because the distribution of the electrons has changed. Transitions between these discrete energy levels occur when radiation is emitted or absorbed, similar to atomic behaviour. There is an important difference, however. If the molecule is in an electronically excited state, the force of attraction between the atoms is different from that in the ground state because the electron distribution in the molecule is changed. Sometimes the attraction between the atoms is weakened to such a large extent that the molecule falls apart. This process can occur for all molecules. For example, OH may be put into an excited state by absorbing radiation, and may then fall apart. This process is called photodissociation. The photodissociation of OH is represented in Fig. 3.4 where OH* designates the unstable intermediate complex. Photodissociation is very often the major means of destroying molecules in astronomical environments. It generally requires radiation in the ultraviolet region of the spectrum. Photodissociation is a process which has no counterpart for atoms. Atoms can be photoionized, and molecules, similarly, may be photoionized, but only molecules may be photodissociated. Molecular and atomic photoionization processes are also depicted in Fig. 3.4.

3.3 Vibration in molecules: more freedom, less choice

The marble in the hollow may oscillate with any amplitude and speed consistent with it remaining in the hollow. However, this is not the case for atoms in molecules; their vibrational and rotational behaviour is controlled by quantum

Photodissociation

OH + radiation → OH* → O + H

Molecular Photoionization

OH + radiation → OH$^+$ + e$^-$

Atomic Photoionization

H + radiation → H$^+$ + e$^-$

Fig. 3.4. Photodissociation and photoionization.

mechanics. The vibrational energy of a system of atoms bound together in a molecule is restricted to certain discrete values, just as is the energy of an electron bound to a proton. For atoms in a hollow of the shape indicated in Fig. 3.5 the energies that are permitted are given by a simple formula

$$E_v = C(v + \tfrac{1}{2}),$$

where C is a constant energy, and v is a whole number or zero, i.e. v can take the values $0, 1, 2, 3, \ldots$, and is called the vibrational quantum number. Quantum mechanics explains why only certain energy levels are permitted: it is because waves associated with the nuclei of the atoms have to fit within the hollow, and these represent the permitted states. Notice that even in the lowest vibrational energy state, with $v = 0$, there is still some energy so the molecule is still

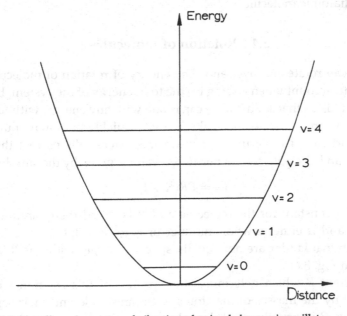

Fig. 3.5. The allowed energies of vibration of a simple harmonic oscillator.

vibrating in the lowest possible vibrational energy state. The molecule never sits entirely still, without vibrating. For the shape indicated in Fig. 3.5 the energy levels given by the formula are equally spaced, by an amount C. The shape, which is called the simple harmonic oscillator potential, is, however, only an approximation to the shape of the hollow in Fig. 3.3. In fact, in a real molecular system the energy levels tend to get closer together towards the top of the well, and there exists a continuum of vibrational energies, corresponding to states in which the separate atoms are free to move far away from one another.

The constant C which determines the spacing of the vibrational energy levels is fixed by the nature of the attractive and repulsive forces in the molecule. The actual value of C is usually much smaller (e.g. in H_2 by a factor of about 20) than the energy associated with electronic jumps. As a consequence, transitions between various vibrational energy levels of the same electronic state leading to the emission or absorption of radiation tend to fall in the infrared, rather than the visual or ultraviolet regions of the spectrum associated with transitions between the electronic states. For example, the molecule CO in its ground electronic state can emit radiation corresponding to the transition from $v = 1$ to $v = 0$ at a wavelength of 4.7 micrometres. Just as in the case of electronic transitions, jumps up and down the vibrational ladder can be brought about by collisions, as well as by the emission and absorption of radiation.

We have described this picture in terms of a molecule with two atoms. However, the vibrations of a molecule with many atoms can be described as a collection of vibrations of this kind, with different constants C. The vibrational energy in a polyatomic molecule would then simply be the sum of terms like that for a diatomic molecule.

3.4   Rotation of molecules

Molecules can rotate end-over-end. The energy of rotation of molecules must be taken into account when adding up the total energy of the system. Everyday experience tells us that solid bodies can rotate with any energy, without restriction. We can throw a stick and make it rotate quickly or slowly. But when it comes to individual molecules, quantum mechanics tells us that the energy of rotation can have only certain permitted values, given by the simple formula

$$E_R = BR(R + 1),$$

where B is a constant for the molecule and R is called the quantum number of rotation and is either a whole number or zero: $R = 0, 1, 2, 3, \ldots$. The steps on the rotational ladder are not equally spaced, but get wider as R increases, as shown in Fig. 3.6.

The constant B which determines the spacing of this rotational ladder is determined by the masses of the atoms in the molecule and their separation. The value of B is usually much less than that of the vibrational constant C

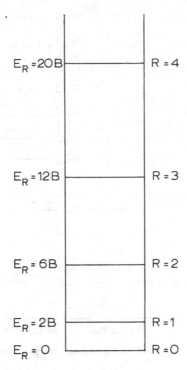

Fig. 3.6. The allowed energies of a rigid rotator.

(e.g. in H_2 and CO by factors of about 70 and 1000, respectively). So when emission by pure rotational jumps occurs within the same electronic and vibrational states, the radiation usually falls in the millimetre to submillimetre region of the spectrum. For example, CO molecules in their ground electronic and vibrational states making the transition from $R = 1$ to $R = 0$ emit radiation with a wavelength of 2.6 mm. Just as for vibration, jumps up and down the rotational ladder can occur through collisions and with the emission and absorption of radiation. For a molecule M in a rotational state with rotational quantum number R the processes are summarized in the equation

$$M(R) \underset{\text{collisions, emission}}{\overset{\text{collisions, absorption}}{\rightleftharpoons}} M(R+1).$$

3.5   Molecular energy levels: getting it together

The total energy of a molecule depends on its particular electronic state, its vibrational state and its rotational state. Each electronic state has a ladder of vibrational levels; each of these vibrational levels has a family of rotational levels associated with it. When transitions occur in molecules, they occur from one particular electronic + vibrational + rotational state to another. The energy of each state is the sum of energies of each part, as shown in Fig. 3.7, and when a photon is emitted it carries away an energy equal to the difference between the energy of the initial state and the energy of the final state.

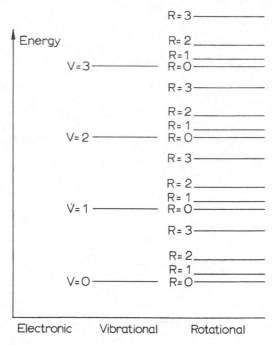

Fig. 3.7. The electronic, vibrational, and rotational energies combined.

In dense enough gas the temperature of the gas determines the population in the various levels. For a molecule in its ground electronic state, collisions with other molecules will, in general, maintain a population in several vibrational levels, and this population will also be distributed over several rotational levels for each vibrational level. Transitions out of the ground electronic state originate on a number of vibrational–rotational levels and populate a number of vibrational–rotational levels in the excited electronic state.

3.6 Introduction to astrochemistry

Our terrestrial experience may give us a false impression about chemistry in the Universe. On Earth, most matter is molecular rather than atomic. The air we breathe is largely molecular nitrogen, N_2, and molecular oxygen, O_2, rather than atomic nitrogen, N, and atomic oxygen, O. Under terrestrial conditions it seems to be easy for atoms to arrange themselves into molecules. However, in almost every extraterrestrial situation conditions are rather different to those on Earth. In many astronomical situations the gas tends to be atomic rather than molecular, and often it is the intense ultraviolet radiation field from very hot stars that is responsible for driving the system towards atoms rather than molecules, by photodissociating the molecules. On Earth, gas densities are high, temperatures are low (but not so low that species react too slowly) and molecules abound. Why is the Earthly chemistry so favoured?

When two atoms approach, the energy of their interaction is something like that shown in Fig. 3.3 for the interaction of two H atoms. If X and Y are free particles to begin with, then this interaction means that they speed up as they move towards one another, then separate, and slow down and unless they have lost some energy they will be able to separate completely. What may happen at the densities typical of the Earth's atmosphere is that while X and Y are interacting, a third atom, Z, collides with the interacting pair, removes some energy and leaves them in a bound state. So high density certainly helps to convert atoms to molecules, because high density leads to frequent collisions. This is the way that many reactions occur when the density is comparable to that of the air. But in nearly all astronomical situations the density is very much lower and the atom Z hardly ever finds the colliding pair XY and so three-body collisions do not contribute to the chemistry. We need to be able to identify some other efficient ways of forming molecules in nonterrestrial situations.

3.7 Ion–molecule reactions: the fast track for chemistry

Reactions between ions and molecules are generally very effective ways of creating 'new' molecules from 'old'. A typical example is the reaction of oxygen ions (oxygen atoms with one negative electron removed) with molecular hydrogen:

$$O^+ + H_2 \rightarrow OH^+ + H.$$

Obviously we need the 'starter' molecule, H_2, but the reaction creates a new molecule, OH^+.

Ion–molecule collisions are very effective in forming new molecules. Reactions occur almost every time an ion and molecule meet, and they are drawn into interaction from relatively large separations. The positive ion polarizes the molecule: i.e. it pulls negative charges slightly towards it. The attraction between these opposite charges is then greater than the repulsion between the positive charges so that a net attractive force results. The partners spiral in towards each other from separations of about ten times a typical atomic radius, and interact with sufficient energy to create a 'complex' of atoms in which the atoms 'forget' from where they have come. In the reaction between O^+ and H_2, the hydrogen atoms forget that they came to the party together. A *ménage à trois* is never a satisfactory basis for a long term relationship, so that – eventually – one pair will decide to stick together and reject the third. In this case, the OH^+ molecule forms, with a stronger bond than H_2, and – as in romances where the deepest affection prevails – the H_2 molecule is lost and an H atom – as the spurned partner – is expelled, carrying off excess energy.

As shown schematically in Fig. 3.8, the reaction of O^+ with H_2 is only the first in a sequence of reactions. Eventually H_3O^+ is formed and it can add no more hydrogen; the oxygen atom is unable to bind any more hydrogen to itself. The efficiency of this chain is high, since reactions occur almost every time a collision occurs.

To make a start in this kind of chemistry we need both a molecule and an ion. In nearly all astronomical situations, the molecule is H_2 and the primary step in

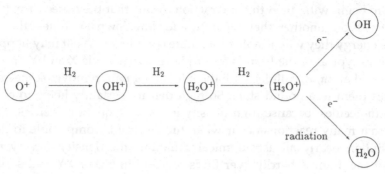

Fig. 3.8. Some key reactions involving oxygen bearing ions.

astrochemistry is usually its formation. *We shall have to consider this step in each of the astronomical scenarios described in the following chapters.* Once H_2 is available, then the effectiveness of ion–molecule chemistry in astronomy is often linked to the rate at which the ions can be created. Ionization can occur in various ways. Ultraviolet radiation with wavelengths of around 100 nanometres can ionize some atoms, e.g.

$$C + \text{radiation} \rightarrow C^+ + e^-.$$

Another possibility is that cosmic rays (c.r.), which are energetic particles (mostly protons) collide with atoms or molecules and eject an electron, e.g.

$$H_2 + \text{c.r.} \rightarrow H_2^+ + e^- + \text{c.r.}$$

Chemically, the most influential cosmic rays in the interstellar gas, star forming regions, and protostellar disks are probably those travelling at about a few tenths of the speed of light. By cosmic ray standards these are not very energetic particles, but they are prevalent, having an abundance of the order of $10^{-4}\,m^{-3}$ in interstellar space.

What will happen to ions such as H_3O^+? They cannot increase in abundance indefinitely. Like any other molecules they may be dissociated by the radiation field. Another important loss mechanism is the reaction with negative electrons. Since the gas is neutral overall, for each positive ion there is a negative electron. The attraction between positive ions and electrons is strong and neutralization usually occurs, but this very often creates an unstable molecule which then falls apart. For instance, as shown in Fig. 3.8, H_3O formed in the recombination of an electron with H_3O^+ dissociates to produce OH or H_2O.

Though this fast dissociative type of recombination is a common fate for *molecular* ions, *atomic* ions, on the other hand, recombine quite slowly with electrons, e.g.

$$C^+ + e^- \rightleftharpoons C^* \rightarrow C + \text{radiation},$$

where the C^* represents a carbon atom that has enough energy in it to separate to an ion and electron, but which – very occasionally – may emit a photon of radiation and stabilize itself.

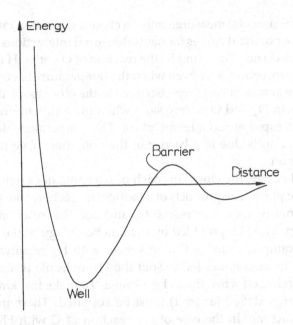

Fig. 3.9. A schematic potential curve showing a barrier as found in the interaction of many neutral–neutral systems.

3.8 Neutral chemistry: change your partners!

When two atoms collide

$$A + B \rightleftharpoons AB^* \rightarrow AB + radiation$$

they are more likely to bounce off each other than to emit radiation and stabilize the molecule AB. What happens if one of the partners is a molecule? Can a new molecule be formed:

$$AB + C \rightarrow A + BC?$$

The answer is: maybe! The situation is not as straightforward as that commonly occurring in ion–molecule reactions. There, the electrical interactions are strong and pull the partners together so that they collide quite energetically, often giving rise to rearrangement. In the case of neutral partners, the force of interaction is weak and there may even be a barrier hindering interaction. It is as if a marble rolling towards a hollow finds a hump in the surface which impedes its progress, as shown in Fig. 3.9. If the barrier is not too great, and the particle has enough energy, the particle may be able to overcome the barrier and reach the well. Then the approach will be close enough that rearrangement may occur, though close approach is often only a necessary and not a sufficient condition for a reaction to proceed, as we describe below.

It is difficult to predict whether a neutral–neutral reaction is hindered by a barrier. In a potential curve like the one shown in Fig. 3.9, a bump that is sufficiently large to retard reactions at the low temperatures (about 10 kelvins) that typify some astronomical environments corresponds to a very small fractional change (about one ten thousandth) in the total potential energy of the complex. Experimental measurements on chemical reaction rates imply that in some cases, the potential curves for neutral–neutral interactions have barriers and in other cases do not. For example, the reaction of O with OH to produce O_2 and H seems to proceed rapidly even when the temperature is very low, so this reaction can have at most a very small barrier. On the other hand, the reaction of N and NO to form N_2 and O is very slow when the temperature is as low as 10 kelvins, but is rapid at room temperature. The temperature dependence of the N + NO reaction is due to a barrier in the potential curve describing the N + NO interaction.

As mentioned above, the close approach of reactants does not ensure that a reaction will take place. The products of a particular reaction may have a higher total internal energy than the reactants, and for the reaction to proceed additional energy must be provided by the kinetic energy of the approaching reactants. For example, to make C atoms react with H_2 requires an input of energy because the energy needed to split the H_2 molecule is actually greater than the energy released when the CH is formed. To make this kind of reaction proceed, the energy deficit (at least) must be supplied. There may also be a barrier to be overcome. In the case of the reaction of C with H_2 to give CH and H there is both an energy deficit and a barrier. Temperatures of several thousand kelvins are required to make this reaction proceed with reasonable efficiency.

Even if there are no barriers, and regardless of whether or not there is an energy deficit, neutral reactions are not in general as efficient as ion–molecule reactions. The charge on the ion means that an ion and a molecule interact significantly at distances much larger than the size of the 'electron cloud' around each reactant. The neutral species, on the other hand, have to experience a close collision before any reaction is possible. Neutral reactions are, therefore, only about 1 per cent as likely to occur as ion–molecule reactions. No simple assumptions can be made about neutral reactions. They need to be studied in detail in the laboratory.

3.9   Heterogeneous catalysis

The final chemical mechanism that we shall discuss is a familiar one: catalysis on surfaces. In astronomy, the surfaces are provided by dust grains (see Chapter 5). The reason why the surfaces of dust grains can be effective in promoting chemistry is straightforward. While atoms A and B might only rarely combine in the gas because the time they spend in collision is so short, on a surface there are sites to which both A and B will be drawn and held long enough for a reaction to occur and for some of the excess energy liberated to be absorbed by the grain. We shall see that dust in interstellar space can provide surfaces on which such catalysis is responsible for the formation of molecular hydrogen (Chapter 5). Molecular hydrogen formed on the surfaces of dust grains can then take part in a variety of gas phase reactions, such as ion–molecule reactions and neutral exchanges.

3.10   Heating and cooling

The temperature of a gas affects the nature of the chemistry that occurs in it. High temperatures mean that collisions are more energetic, so that neutral exchange reactions that are inhibited at low temperatures can occur at high temperatures. The temperature also affects in a sensitive way the radiation that is emitted by the gas. In a warm gas, collisions between molecules help to populate a wider range of rotational and vibrational levels than is possible in a cold gas. The temperature is clearly an important parameter in any astronomical situation. We shall discuss in the succeeding chapters the processes that control the temperature in many of the different astronomical environments. Here, we merely mention a few general considerations. Heating and cooling of astronomical gases occur by a variety of processes, with those involving radiation often being the dominant ones in astronomical sources.

Heating can occur when matter extracts energy from the radiation field. For example, if ultraviolet radiation (perhaps from a hot star) ionizes a hydrogen atom then the electron carries off that amount of energy that is in excess of the minimum energy required to release the electron from the proton. In subsequent collisions, the electron shares its energy with the gas and loses energy itself in the process, while the gas of H atoms gains in energy, i.e. becomes hotter. In this way, energy produced by thermonuclear processes deep within a star and emerging as ultraviolet radiation heats the gas surrounding the stars.

Cosmic rays are also energy sources, and are particularly important in regions where stellar ultraviolet radiation does not penetrate. As mentioned in Section 3.7, cosmic rays colliding with atoms or molecules ionize them, producing electrons which carry off some energy which is then transferred to the neutral gas via collisions.

Cooling of a gas generally occurs by the emission of radiation from atoms and molecules. Consider, for example, a gas of molecular hydrogen. If this gas is sufficiently warm, then collisions may make a particular H_2 molecule jump into the first excited vibrational state ($v = 1$)

$$H_2(v = 0) + H_2(v = 0) \rightarrow H_2(v = 0) + H_2(v = 1)$$

and the energy needed for this jump has come from the energy of motion of the H_2 molecules relative to one another. After an interval, the excited H_2 molecule may radiate

$$H_2(v = 1) \rightarrow H_2(v = 0) + \text{radiation}$$

and the radiation escapes from the gas. Therefore, energy that was in the gas has now escaped, and so the gas cools. There are many ways in which cooling can occur, involving a wide variety of atomic and molecular excited levels. The temperature in a gas is then a result of a balance between the heating and cooling mechanisms.

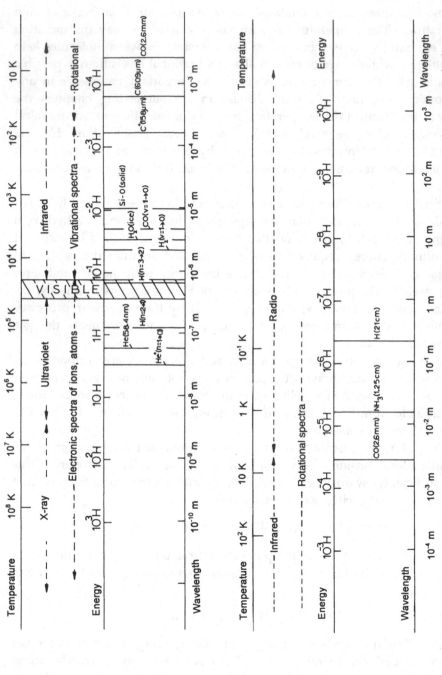

Fig. 3.10. The electromagnetic spectrum. The diagram indicates energies (measured in units of H, the energy required to ionize the hydrogen atom) and wavelengths (in metres) corresponding to temperatures (measured in kelvins, denoted K). Some astronomically important atomic or molecular lines are indicated (with wavelengths given, as appropriate, in nanometres (nm), micrometres (μm), millimetres (mm) and centimetres (cm)).

3.11 The electromagnetic spectrum

We include Fig. 3.10 to show the extent of the electromagnetic spectrum. The visible part of the spectrum, to which our eyes respond, is a very small part of the entire range, and runs from about 350 to 700 nanometres. At wavelengths shortward of the visible part is first the ultraviolet region, then the X-ray region. At wavelengths longer than optical wavelengths lie the infrared and radio regions of the spectrum. As well as giving the wavelengths, Fig. 3.10 also indicates the energies at which some astronomical important atomic or molecular transitions occur. Since the energy is proportional to frequency, it is inversely proportional to wavelength: that is, long wavelengths mean low energies, and *vice versa*. The unit of energy adopted here is the minimum energy required to ionize a hydrogen atom (13.6 electron-volts or 2.18×10^{-19} joules; this is the energy A of Fig. 3.1). One unit of energy then corresponds to a wavelength of radiation of 91.2 nanometres. With the energy scale in place, we can see that different types of transition fall in different regions. At the highest energies (shortest wavelengths) fall the electronic spectra of ions, atoms, and molecules. In the infrared we find wavelengths of radiation corresponding to vibrational transitions of molecules, while at longer wavelengths, in the radio, lie lines associated with rotational transitions of molecules. The diagram indicates a few lines of each type corresponding to astrophysically important transitions.

Since temperature is a measure of the average kinetic energy of a particle in a gas, we show also the temperature corresponding to the energy of the transition. This temperature actually overestimates slightly the temperature at which matter can be expected to radiate at the wavelength indicated. Special processes may, however, affect individual lines, so this temperature indication is no more than a guideline. We see, for example, that the visible region corresponds to temperatures of the order of 10 000 kelvins while gas at a million kelvins or so can be expected to radiate in the X-ray region. The cool matter in the Universe, such as the Earth at a few hundred kelvins above absolute zero, radiates in the infrared.

Selected references

Duley, W W and Williams, D A: *Interstellar Chemistry*, Academic Press, London (1984).
Hartquist, T W (ed.): *Molecular Astrophysics – A Volume Honouring Alexander Dalgarno*, Cambridge University Press, Cambridge (1990).
Herzberg, G: *Molecular Spectra and Molecular Structure – I. Spectra of Diatomic Molecules*, Van Nostrand Reinhold Company, New York (1950).
Landau, L D and Lifshitz, E M: *Quantum Mechanics*, Pergamon Press, Oxford (1965).

4

Chemistry after the Big Bang

___

In India, for example, where the first form to appear in the lotus of Vishnu's dream is seen as Brahma, it is held that when the cosmic dream dissolves, after 100 Brahma years, its Brahma too will disappear – to reappear, however, when the lotus again unfolds. Now one Brahma year is reckoned as 360 Brahma days and nights, each night and each day consisting of 12 000 000 divine years. But each divine year in turn consists of 360 human years; so that 'one full day and night of Brahma, or 24 000 000 divine years, contains 24 000 000 times 360 or 8 640 000 000 human years . . .'

Every day of a Brahma lifetime of 100 Brahma years, the god's eyes slowly open and close 1000 times. When they open a universe appears, and the moment they close it fades.

Joseph Campbell in *The Mythic Image*

In the conventional view, the Universe began in a hot Big Bang some 15 billion years ago, and has been expanding ever since. At a very early stage in the expansion, when the temperature of the Universe was still some hundreds or thousands of millions of kelvins, collisions between subatomic particles created hydrogen and helium nuclei with very minor traces of deuterium (or heavy hydrogen), lithium, beryllium, and boron nuclei. The matter was almost entirely ionized at this stage. However, as the expansion continued, the wavelengths of the radiation filling the whole Universe (and left over from the Big Bang) continually became longer, in direct proportion to the degree that the Universe had expanded; most of the radiation consequently acquired wavelengths longer than those necessary for it to ionize hydrogen and helium. Today, this cosmic background radiation has a 'black body spectrum' that peaks at a wavelength of about 2 millimetres. The discovery in 1965 and study of this radiation provides support for the Big Bang model of the Universe and constraints on models of galaxy formation. The gas temperature fell to several thousand kelvins, matter began to recombine, and the Universe became more neutral as the expansion continued.

Due to their self-gravity, certain regions of matter within the generally expanding Universe collapsed to form protogalaxies. By protogalaxy we mean a gaseous object collapsing to form a galaxy but that has not yet fragmented sufficiently for stars to form within it. The only force capable of causing the collapse of matter over large regions of space is the force of gravity. But how did gravity select certain regions? Why did they become unstable? It now seems clear that a rudimentary chemistry played a fundamental role by permitting gravitational collapse, once started, to continue. Therefore, even before our own Galaxy was formed, even before the elements of oxygen, carbon, nitrogen, and iron – the stuff of planets and of life itself – were created in stars, a simple chemistry was in action and was influencing the evolution of the Universe by permitting galaxies to form.

The story of chemistry in the Early Universe was not completed with the formation of protogalaxies. Once galaxies had formed and stars were born in them, they began to exert an influence on their surroundings. Some energetic galaxies generated powerful winds which – in impacting on the intergalactic gas – created conditions under which the same simple chemistry proceeded more efficiently. The presence of molecular hydrogen, formed in this chemistry, helped to produce from the impacted gas some subgalactic units of matter which collapsed and had an independent existence. Perhaps some of those subgalactic objects became trapped by pre-existing galaxies; alternatively, maybe groups of them coalesced to form younger galaxies.

4.1 The first chemistry in the Universe

The gas in the Early Universe was mostly atomic hydrogen. Helium was also present, with an abundance about one tenth that of hydrogen. Deuterium (D, or ^{2}H) was less abundant than hydrogen (the main isotope, ^{1}H), and was present at a level of about one part in 100 000. Lithium, beryllium, and boron were also present, but they were even less abundant. Although the gas was initially ionized, as the expansion proceeded the gas cooled and the protons and electrons recombined radiatively

$$H^+ + e^- \rightarrow H + radiation,$$

where the radiation was emitted in various ways involving several steps, so that the photons of the radiation did not have energy enough to ionize other hydrogen atoms. Although the tendency was towards neutrality, the universal expansion ensured that not all the electrons and protons could find one another, i.e. the Universe continuously became a bigger place (and this expansion still continues) so collisions became less frequent. The recombination of electrons and protons was, therefore, never quite complete and the gas remained slightly ionized; this is a point of crucial importance for chemistry. The temperature of the gas was about 3000 kelvins when most of the recombination was completed. We shall discuss how the gas cooled from this temperature in Section 4.3, but, initially, the cooling was simply due to the expansion of the Universe.

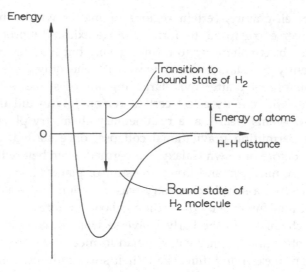

Fig. 4.1. Energy loss required for H_2 formation. The lowest H—H potential energy curve (the one for H_2 in its lowest electronic state) is shown. The energy of two approaching unbound H atoms remains constant as the atoms become closer. Consequently the H atoms have too much energy to be bound by the attractive interaction described by the potential curve, and though they may approach one another, they will fly apart again. Indirect chemical ways of removing enough energy from two H atoms to allow the bound state of H_2 to form are described in the text.

What kind of chemistry can there be in a gas of H atoms, when the gas is largely neutral and fairly warm? If two H atoms could associate directly to form a hydrogen molecule, H_2, merely by colliding, then that would be a most efficient chemistry. However, for H atoms in their ground states, this is impossible. The atoms merely bounce off each other, and transfer between themselves some energy and momentum. To stick to each other and form a molecule requires that they – as a system – lose some energy while they are in contact; then they would have insufficient energy to separate from each other and a molecule would be formed, as illustrated in Fig. 4.1. On the Earth, densities in gases are so great that while a pair of atoms are inter-acting, a third atom collides with them and removes some energy from the pair, leaving them bound together (see Section 3.6). But densities in the Early Universe were much too low for this to occur, so three-body collisions played no role there.

Atoms can sometimes associate radiatively to form a molecule. The radiation emitted by the colliding pair carries energy away from the atoms, leaving them with insufficient energy to separate again:

$$A + B \rightarrow AB + \text{radiation}$$

but for H atoms in their ground states this does not occur. If one of the H atoms is in an excited state, radiative association can and sometimes will proceed, but in the Early Universe the populations in the excited states were so small that

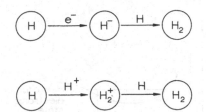

Fig. 4.2. Schematic diagrams showing chemical schemes by which H_2 formed in the Early Universe.

this process was insignificant. Any chemistry in the Early Universe must have been more complicated than this kind of direct association. There are two indirect routes which allow H_2 to form; they are illustrated in Fig. 4.2.

The first involves electrons: these can attach themselves to H atoms to form the negative ions of the species, H^-, as radiation is emitted. This is rather inefficient and occurs in about only one in a million collisions. But there was plenty of time in the Early Universe, and the gas was mostly hydrogen, so many collisions of electrons with H atoms did occur. The H^- ions were then more likely to collide with H atoms than with any other species, and when they did H_2 formed quite efficiently. As the H_2 formed from H^- and H an electron became unbound and carried off energy, leaving the H_2 bound. The electron was then available for another reaction forming H^-. The electron, therefore, behaved as a catalyst, enabling the association of two H atoms to occur indirectly. Of course, the two-step sequence starting with the formation of H^- could be interrupted; the weakly bound electron of the H^- ion was often detached by long wavelength radiation or removed when H^- collided with a proton.

The second route to H_2 production involved protons instead of electrons. First, protons and H atoms associated to form H_2^+ which became bound because radiation was emitted and carried off energy. Then the H_2^+ reacted with H to form H_2 and H^+. In this sequence, the proton acted as a catalyst. This sequence of reactions could also have been interrupted, since H_2^+ could be dissociated by radiation or neutralized by capturing an electron leading to the production of two neutral H atoms.

Of the two catalysts, electrons were the more effective, but even they were not very efficient. The H_2 that was formed could be destroyed by radiation in a two-step process depicted in Fig. 4.3, and by energetic collisions with H where the energy of the collision was great enough to tear the atoms apart. The balance of formation and destruction gave a small H_2 density; this was about one or two molecules of H_2 per million H atoms. However, even this low level of abundance had important consequences.

Deuterium behaves chemically as hydrogen, so the reactions involving H in the Early Universe also applied to D. We expect the molecule HD to have been present at an abundance of about one molecule of HD to 100 thousand of H_2.

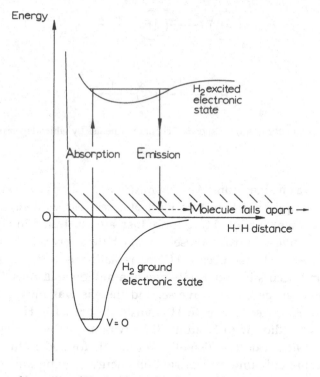

Fig. 4.3. Photodestruction of H_2. An ultraviolet photon is absorbed by H_2 in its ground electronic state producing H_2 in an excited electronic state. The two H atoms remain bound to one another in this excited electronic state but it eventually is depopulated by the emission of radiation. In about 10 per cent of these transitions the emitted radiation does not carry away enough energy for the H_2 to return to a bound vibrational level in the ground electronic state. Rather, the vibrational continuum of the ground electronic state is populated, and the two H atoms are no longer bound together and fly apart.

4.2 Cooling and collapse: the consequences of chemistry

As gravity caused a protogalaxy to collapse, much of the gravitational energy released in the collapse was converted to heat in the gas. As we discuss more fully later in this section, the pressure of hot gas retards the collapse of objects. In order for galaxies to form, cooling mechanisms had to operate to counteract the input of heat due to the collapse. Any gas at tens of thousands of kelvins in the Early Universe after recombination cooled rapidly by converting energy of the atoms' motions into radiation. It did this in the following way: two H atoms collided with sufficient energy to ionize at least one of them

$$H + H \rightarrow H + H^+ + e^-$$

so that some energy from the gas was used in ionizing and some remained as energy of motion in the products. When an electron and a proton later recombined the energy with which the electron and proton interacted was radiated away in several steps. The net result was that energy from the gas was

36

converted to radiation which could not be absorbed by other atoms. The gas was therefore less energetic, and, therefore, cooler, than before.

Cooling was also caused by the collisions of the electrons with neutral H atoms. A collision inducing the excitation of an H atom to its $n = 2$ electronic state removed thermal energy from the gas, and eventually the atom radiated and returned to its ground state.

The gas cooled in these ways quite efficiently when there were many collisions with energies capable of ionizing hydrogen. But as the cooling continued to temperatures below about 10 000 kelvins, fewer and fewer collisions met this energy requirement and the rate of cooling declined. In regions of gas where collapse began, the temperature would have remained at roughly 10 000 kelvins if cooling processes other than these mechanisms had not been involved.

Molecular hydrogen formation introduced new cooling mechanisms into the Early Universe, and these allowed cooling to continue to much lower temperatures, around 100 kelvins, as collapse to high density occurred. This cooling enhanced the ability of regions to collapse. As we saw in Section 3.10, collisions between H atoms and H_2 molecules in a warm gas can raise the molecules to excited vibrational levels. These vibrational excitation collisions are efficient at temperatures of several thousand kelvins, or higher. Such collisions in the Early Universe converted energy from the gas to vibrational energy in the molecule. The molecule then jumped back down to the ground vibrational level and emitted the energy as radiation so that the gas lost energy, and therefore cooled. The H_2 molecule was then available to be used again as a coolant. In the Early Universe when the temperatures were several thousand kelvins and above, the vibrational levels of H_2 were readily populated by collisions, so H_2 was an effective coolant at these temperatures.

Hydrogen molecules can also rotate (see Section 3.4), and collisions with H atoms can make the molecules rotate faster. Again, just as for vibrationally excited molecules, a rotationally excited molecule can jump to some lower state of rotation by emitting radiation, and if this happens the gas becomes cooler. In a gas at temperatures above about 100 kelvins many of the collisions are energetic enough to populate excited rotational levels of H_2; H_2 can provide cooling over a wide range of temperature, and in the Early Universe was an important coolant.

The HD molecule is actually more efficient, per molecule, as a coolant at very low temperatures. Even though it was at a very low abundance in the Early Universe, its contribution to cooling was also important, particularly when the temperature was below about 100 kelvins, since its lowest easily excited rotational level is at a much lower energy than that of H_2.

Let us now consider the consequences of the ability of the gas to cool. If gravity is to cause a region to collapse, then the gravitational attraction must pull inwards harder than the gas pressure pushes outwards. As the collapse proceeds, the pressure increases, and if the gas is tenuous the gravitational forces are weak; therefore, tenuous, hot gases are not likely to collapse. In fact, spherical regions of gas must have a mass greater than a critical value, called the Jeans mass, M_J, if gravity is to overcome internal pressure. The Jeans mass is approximately

$$M_J = 2 \times 10^4 (T^3/n)^{1/2} \text{ solar masses,}$$

where T is the temperature in kelvins, n is the number of H atoms per cubic metre (n is called the number density), and the mass is calculated in solar masses (about 2×10^{30} kg). The important point to note from this formula is that the Jeans mass is larger at higher temperatures, but smaller at higher densities.

The sizes of the first regions to collapse in the Early Universe are unknown. Indeed, this has been a topic of much debate since the mid-1960s. Electrons and protons recombined so that the Universe was largely neutral atomic hydrogen when the Universe was about 5×10^5 years old. The temperature was then about 3000 kelvins and the number density, n, was about 10^{10} m^{-3}, corresponding to a Jeans mass of about 10^5 solar masses. However, interaction between radiation and matter before recombination had occurred created a kind of friction that smoothed out any fluctuations with masses less than about 10^{13} solar masses. (The Milky Way mass is a few per cent of this.) Possibly, giant fluctuations of this size or greater were the first to collapse, and fragmentation in them created the protogalaxies in which further fragmentation created the stars. However, we do not know how this hierarchical fragmentation took place. In fact, many cosmologists suspect that fragments of the sizes of galaxies or smaller were well developed long before these giant fluctuations had collapsed very far. A major problem is that for fragmentation to occur the fragment must be more massive than the Jeans mass. This problem is amplified because during the collapse of a fragment, energy must be radiated away or the gas will get so hot that the Jeans mass will increase up to the fragment's mass, terminating the collapse. Clearly, a great deal of cooling had to occur to allow the Jeans mass in collapsing protogalaxies to fall from its value of 10^5 solar masses at the time of recombination to a value of about 1 solar mass so that stars could form. In fact, it is likely that collapse really got going when the average number density of the Universe had dropped to 10^4 m^{-3}, at which time the Universe was about 5×10^8 years old and the gas had a temperature of somewhat less than a kelvin. Collapse to the average density in the Galactic disk of about 10^8 m^{-3} (including matter in stars) would have increased the temperature to hundreds of kelvins while collapse to stellar densities would have taken the temperature higher.

There had to be a means of removing energy from the gas as the collapse proceeded to form the first stars. In fact, emission of radiation by H_2 provided this cooling in the Early Universe. As we have seen, this H_2 cooling mechanism operates over a range of temperature from several thousand kelvins down to about 100 kelvins. The cooling rate, even with the low H_2 abundance of one molecule in a million H atoms, was substantial and sufficient to remove the heat content of the gas in the collapse time, so that the gas never heated up.

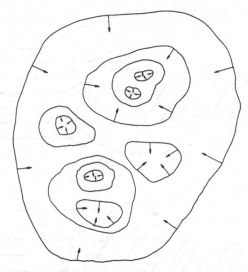

Fig. 4.4. The hierarchy of collapse at constant temperature. If the cooling processes are efficient enough to keep the temperature constant, the Jeans mass (the minimum mass of an object that can collapse independently) decreases as collapse occurs. Thus, fragmentation to smaller and smaller objects can occur.

Therefore, a collapse, once started, continued, and the Jeans mass became smaller and smaller, because the temperature did not increase while the density increased. This means that during collapse less and less massive units became gravitationally unstable, as depicted in Fig. 4.4. How these subgalactic units aggregated to form a protogalaxy is not yet well understood. But the significance of H_2, and the chemistry that formed it in the Early Universe, is clearly evident.

4.3 Intergalactic shocks

During their adolescence some galaxies were violent objects. Some developed fast winds of gas powered by young stars and possibly by black holes as much as a billion times as massive as the Sun. These winds impinged on the surrounding gas. The speeds of the winds were many hundreds of kilometres per second and highly supersonic. A wind carried many tens of solar masses per year out of the most active galaxies.

Before we consider the effect of a supersonic wind on the surrounding gas, let us consider a more familiar example of supersonic flow. A plane flying through the air experiences the same flow of gas around it as it would if it were at rest and the air flowed past. Figure 4.5(a) shows the flow of gas around a plane from the perspective of an observer in another plane flying parallel to it. When the wind speed is moderate, the air ahead of the plane begins to adjust to the obstacle (i.e. the plane) upstream, and the flow lines in the gas stream smoothly around the plane. Information to the incoming gas about the obstacle is transmitted upstream at the speed of sound in the gas, so the gas 'knows' that it must avoid the plane.

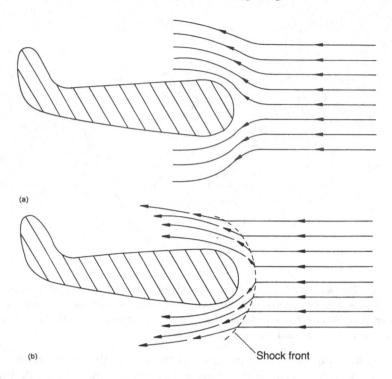

(a)

(b)   Shock front

Fig. 4.5. A flow around a plane. (a) The upstream flow can adjust since information about the plane's shape and position is transmitted at the sound speed which is greater than the upstream wind speed as measured from the plane. (b) The upstream flow is faster than the sound speed. Information of the plane's position and shape cannot propagate upstream, leading to a shock developing around the plane.

However, when the speed of the flow is supersonic (i.e. faster than the sound speed) there is no way that information about the presence of the plane can reach the upstream gas. The flow lines for the stream are unperturbed, until – rather near the plane – a shock front develops around its head (see Fig. 4.5(b)). The flow velocity changes abruptly at the shock front, and most of the energy that was in the bulk motion of the gas is transformed to heat. The shocked gas moves more slowly around the plane. The effect of a shock is therefore to change quite abruptly the character of the gas, i.e. its density, temperature, and velocity. This is illustrated in Fig. 4.6. Magnetic fields may sometimes help to smooth these abrupt changes, but we shall not be concerned with those effects here.

The fast wind from a galaxy feels the intergalactic gas as an obstacle which must bring the flow nearly to a halt. Since the wind speed is supersonic, the deceleration is effected by a shock, and after the wind is shocked it is very hot indeed, probably more than 10 million kelvins. The unshocked and shocked wind occupies regions represented in Fig. 4.7 by locations (a) and (b), respectively. The wind drives a bubble that must expand because the temperature and pressure of the shocked wind are so great. It expands against the cool intergalactic gas, and the speed of this expansion is

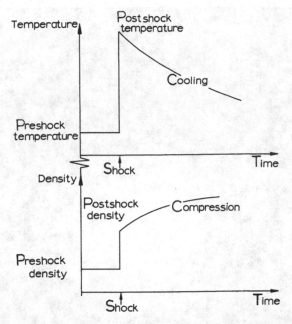

Fig. 4.6. The temporal history of the physical properties of a parcel of gas passing through a shock. When the shock is encountered the parcel is heated, slowed, and compressed. Cooling due to collisionally induced excitation followed by radiation takes place and, because the pressure remains relatively constant, leads to further compression.

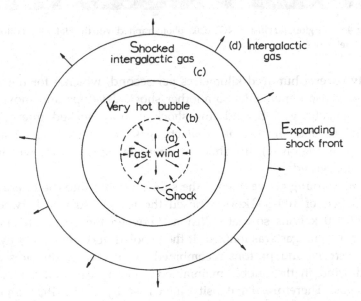

Fig. 4.7. The wind-blown bubble around a young galaxy. See the text for a full explanation of the structure.

Fig. 4.8. The globular cluster, NGC 6205, photographed with the Hale Observatory 200 inch telescope.

probably several hundred kilometres per second, which – for the intergalactic gas – is also supersonic. So another shock develops and moves out into the intergalactic gas. Behind this shock is hot shocked intergalactic gas. The ambient and shocked intergalactic gas are in regions (c) and (d) in Fig. 4.7. It is region (c), the shocked intergalactic gas, with which we shall now be concerned.

Around a young active galaxy, the shock passing into the intergalactic gas with a speed of 100–300 km s^{-1} raised the temperature to between 100 000 and 1 000 000 kelvins so that collisions between the atoms stripped off the electrons, i.e. the gas was ionized. It then cooled very effectively by radiating as the electrons and protons recombined. As it cooled, the pressure of hot shocked wind in the bubble maintained a nearly constant pressure on the cooling gas. Therefore, its density increased by about 1000 times from a value of about 10^3 m^{-3} in the ambient intergalactic gas to a value of about 10^6 m^{-3} at a temperature of around 10 000 kelvins.

In these relatively high density circumstances, and because the ionization remained somewhat elevated longer than the cooling to 10 000 kelvins required, H_2 began to form, and a much higher fraction of gas was converted to H_2 here than in the unshocked pregalactic gas. The abundance of H_2 molecules was about one for every thousand H atoms in the swept-up intergalactic gas, a thousand times greater than in the ambient intergalactic gas. The H_2 cooling effects were consequently much more powerful, densities increased further, temperatures were low and the Jeans masses were relatively small. In the cooled postshock intergalactic gas the temperature was probably as low as 100 kelvins, while the density was about 10^6 m^{-3}. Consequently, the Jeans mass was about 10^5 solar masses, which is much less than the mass of a galaxy. This was most likely the typical fragmentation mass in the second generation of gravitational collapse and may be the mass of the basic units from which the *second* generation of galaxies formed. The mass of 10^5 solar masses is characteristic of collections of stars called *globular clusters*, one of which is pictured in Fig. 4.8. It has been conjectured that at least some galaxies form by an agglomeration of globular clusters, and – possibly – other galaxies grew by accreting them. Of course, if a young galaxy had, in addition to a powerful wind, an intense source of radiation, then the abundance of molecular hydrogen would have been affected.

Through H_2, therefore, the size of the subgalactic units was determined. The formation of H_2, and the consequent cooling properties determined how the Universe became a collection of galaxies.

Selected references

Dalgarno, A and Lepp, S: 'Chemistry in the Early Universe', in *Astrochemistry – IAU Symposium No. 120*, eds. Vardya, M S and Tarafdar, S P, D Reidel Publishing Company, Dordrecht (1987).

Shapiro, P R: 'Chemistry in the Early Universe', in *Astrochemistry of Cosmic Phenomena – IAU Symposium No. 150*, ed. Singh, P D, Kluwer Academic Publishers, Dordrecht (1991).

5

Interstellar clouds – the birthplaces of stars

In the previous chapter we described the influence of chemistry on the Universe before and as the galaxies formed. In the rest of this book, except for Chapter 11, we shall concentrate on aspects of our own galaxy, The Milky Way, or, simply, the Galaxy. In this chapter we shall consider the role of chemistry in the interstellar medium, the material that fills the space between the stars of the Galaxy. It is a medium consisting of a large variety of types of diffuse gas from tenuous (about 10^3 atoms, ions, and electrons per m^3), hot (about one million kelvins) plasma to cold (about 10 kelvins), dense (about 10^9 atoms and molecules m^{-3}) clouds some of which contain even denser regions in which stars are being formed.

A rich chemistry is present in many of the clouds of interstellar space. The molecules are obviously convenient tracers of the gas, but do they play any other role? What is the chemistry by which these molecules form? We shall address these questions and also be concerned with the ways in which chemistry affects the thermal properties of the clouds. We shall consider the way in which the magnetic field retards the collapse of a cloud to form a region in which stars are born and the chemistry that establishes the ionization properties which, in turn, govern the magnetic properties of the clouds. The means by which we infer the rate at which cosmic rays induce the ionization of H_2 (an important step in the chemistry controlling the ionization properties in the denser, darker clouds) will be described, as will the ways that molecules in present day interstellar clouds are used to learn more about the Big Bang which occurred about 15 billion years ago.

5.1 The Milky Way and its nonstellar content

Galaxies are found to possess a wide range of shapes and physical character-istics. Some are irregular, apparently chaotic assemblies of matter. Others are more organized in appearance. The Milky Way is a disk-shaped collection of stars, gas, and dust. It is remarkably flat and thin; it rotates and has spiral structure. The diameter of the galactic disk is about 100 000 light years, and its thickness is about 6000 light years. The Milky Way contains some 100 billion (10^{11}) stars and is similar in size and morphology to many other galaxies that we can conveniently study, such as NGC5194 which is pictured in Fig. 5.1. The Sun, a typical low-mass star of modest intensity, is situated near the central plane of the disk of the Galaxy, roughly twice as far from the Galactic Centre as from the outer edge of the distribution of fairly luminous stars.

Spiral galaxies similar to the Milky Way are numerous, and are oriented in many different planes. NGC5194 is almost exactly face-on to the line of sight from the Earth, and we can clearly see the spiral structure. Sometimes, however, the line-of-sight is in the same plane as the disk of the observed galaxy and we see it edge-on. Such a galaxy is pictured in Fig. 5.2. A dark lane, rather more tightly confined than the distribution of stars is clearly apparent around the central plane of the galaxy in Fig. 5.2. This dark lane is direct evidence of the presence of interstellar matter. The dust in clouds of gas and dust near the planes of this and other spiral galaxies obscures the light of stars, giving rise to apparently dark structures. We shall describe the properties of dust in Section 5.2, but here we note that gas and dust are gener-ally quite well mixed, so that the presence of dust, as indicated by obscuration of starlight, implies the presence of gas.

The gas in other galaxies, and in our own, can be traced directly. In Fig. 5.3 contours of emission at 21 cm from hydrogen atoms are overlaid on an optical photograph of a spiral galaxy. The gas is seen to be broadly coincident with the spiral arms. The 21 cm emission arises in regions that are primarily atomic, but

Fig. 5.1. A photograph of a spiral galaxy similar to the Milky Way. The galaxy is the 'Whirlpool', NGC5194, and has a satellite galaxy, NGC5195. The photograph was taken at the Hale Observatory with the 200 inch telescope.

a map of CO emission, which arises in gas that is mostly molecular, would show that it too is strongest in the spiral arm regions.

5.2   Interstellar dust: soot and sand

As we discussed in Chapter 2, once a galaxy has formed and stars begin to burn their fuel, i.e. hydrogen, then the ashes of those thermonuclear processes begin to accumulate. These ashes, principally in the form of carbon, nitrogen, and oxygen are ultimately returned to interstellar space in stellar winds or in supernova explosions. There, they are incorporated in the gas which forms the next

Fig. 5.2. A galaxy similar to the Milky Way towards which the line of sight lies in the same plane as its disk. The galaxy is NGC4565, photographed with the 200 inch telescope at the Hale Observatory.

generation of stars. The Earth and all its inhabitants are made of these ashes. The atoms of which each of us is made were at one time in the interior of a star more massive than the Sun.

The presence of these elements obviously makes possible a much richer chemistry than that which we described for the Early Universe. These elements also permit the existence of interstellar dust, the darkening effects of which are so apparent in Fig. 5.2. Many cool stars (with stellar envelope temperatures of about 2000 kelvins) produce particles of dust in their atmospheres, and blow these particles into space. (The chemistry in the envelopes of these stars is discussed in Chapter 8; chemistry is important for the production of dust.) Cool carbon-rich stars make sooty particles, much as a candle has a sooty flame. Some of these stars are seen to pulsate, as dust forms and is blown away. Cool oxygen-rich stars make long lived silicate dust and this, too, is blown into the surrounding medium. These dust components, and others formed in novae and supernovae explosions, are thoroughly mixed with the interstellar gas. Turbulence in the gas stirs the mixture as a cook's spoon mixes raisins in a cake mix. After stirring it is difficult to separate the raisins from the dough; so it is with gas and dust, which generally remain well-mixed.

Just as the absorption of light by dust in other galaxies provides information about the interstellar media in them, the 'fog' caused by dust in the interstellar medium of the Milky Way can be used to infer much about its interstellar matter. Many optical photographs of rich star fields in the Galaxy show dark regions where stars seem to be much fewer in number than usual. Figure 5.4 shows a photograph of such a region. A deeper study of such regions shows that the background stars are not absent but simply heavily obscured. Counts of the stars can be employed to deduce that the dust distribution is extremely clumpy with nearly all the dust occupying only about 1 per cent or so of the interstellar volume. The dusty clumps are called clouds. We shall treat the natures of clouds more fully in the next section; here we concentrate on the dust.

The obscuration of a star by a dusty cloud is always more pronounced for blue light than for red; this differential extinction makes all stars seen through a dust cloud appear somewhat redder than they should, just as the Sun when viewed near the horizon through the Earth's dusty atmosphere is also perceptibly reddened. The general trend of increasing extinction of starlight by dust towards shorter wavelengths extends from the infrared right into the ultraviolet. Figure 5.5 shows how the extinction varies with wavelength. Using the information in this figure, and making reasonable estimates of the optical properties of dust allows us to deduce the basic parameters of interstellar dust. We find that dust comprises about 1 per cent by mass of the interstellar medium. The dust is made up (mainly) of particles of carbon and of silicates, with dust particle diameters comparable to the wavelength of visible light (about 500 nanometres) and extending to quite small sizes (less than 10 nanometres). Small particles are much more numerous than large though altogether they contain somewhat less mass than all of the large particles.

Interstellar dust has several quite fundamental properties which play a crucial role in making our Galaxy as it is. Absorbing and scattering starlight, the dust reduces or eliminates the intense destructive power of starlight. The radiation absorbed by the dust has two important effects. Firstly, it causes electrons to be ejected from the dust grains, and these are an important source of energy of the gas; they take energy from starlight and share it with the gas. Secondly, the radiation also tends to heat the dust grains, so that they radiate in the infrared region. Dust grains therefore transmute radiation from the ultraviolet and visible into the infrared.

Dust grains also have an important chemical role: they provide surfaces on which hydrogen atoms can recombine to form H_2 molecules. This is an effect commonly seen in the laboratory; indeed, it is difficult to stop an atomic gas

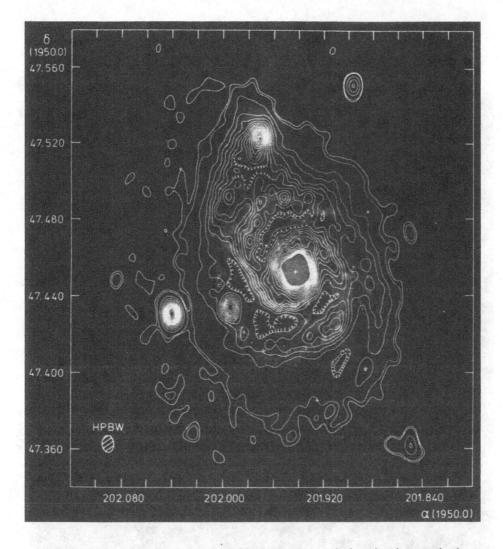

Fig. 5.3. Contours of hydrogen atom 21 cm emission overlaid on the photograph of a spiral galaxy NGC5194. The intensity of the hydrogen emission at a point is proportional to the number of contours that enclose that point. (From A Segalovitz in *Astronomy and Astrophysics*, vol. 54, p.703 (1977).)

from recombining into molecules by reactions at the surface of the flask which contains it. The reaction to form H_2 on interstellar dust is the initial step in all of interstellar chemistry (see Section 3.9).

At the centres of thick clouds, dust grains are generally fairly cool (less than about 10 kelvins) and protected from radiation that might influence the chemistry taking place at their surfaces. This means that molecules such as H_2O stick with high efficiency, forming icy mantles. Such mantles are positively identified by spectroscopy. The loss of molecules from the gas by incorporation into solid ices can have a significant effect on the chemistry in an evolving molecular cloud (see Chapter 6).

Fig. 5.4. A photograph of a Galactic region with few easily seen stars. The region with few stars is a dark globule of gas and dust that extinguishes the light of stars behind it. The globule is illuminated by the star at one side of it. (From: P Murdin and D Allen, *Catalogue of the Universe* (photographs by D Malin), Cambridge University Press, Cambridge, 1979.)

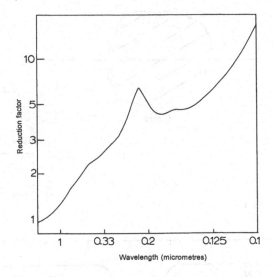

Fig. 5.5. The dependence of interstellar extinction on wavelength. The extinction along a path of 1000 light years through an interstellar medium with an average number density of hydrogen nuclei of $10^6\,\mathrm{m}^{-3}$ is shown. The vertical axis gives the factor by which the background starlight is extinguished. The horizontal axis runs from infrared wavelengths to visual wavelengths (0.35–0.7 micrometres) into the ultraviolet. At an ultraviolet wavelength of 0.1 micrometre the intensity of the radiation field is reduced, due to extinction, by a factor of more than 10.

5.3   Direct observations of the gas in the different types of interstellar cloud

Though the cloudiness of the interstellar medium can be inferred from the observations of the obscuration of stars, direct spectroscopic studies of the interstellar gas yield the most precise information about it. Optical absorption by atomic species along lines of sight towards bright stars provided the earliest data about the densities and temperatures of interstellar gas that is not so close to any one star as to be unduly influenced by it alone. The principle on which absorption studies are based is illustrated in Fig. 5.6(a); in the absence of intervening gas no trough in the spectrum would be observed. A different type of datum useful in the analysis of interstellar gas became obtainable in the 1950s and 1960s when radio telescopes were used extensively to map the Galactic distribution of 21 cm emission from atomic hydrogen.

The dust distribution, optical atomic absorption, and 21 cm data were employed in the development of a picture of the interstellar clouds. Clouds through which around a quarter of the optical starlight was absorbed were found to be prevalent with one occurring on average roughly every 500 light years along a line of sight. The number density of hydrogen atoms, the temperature, and extent of such a cloud were found to be about $3 \times 10^7\,\mathrm{m}^{-3}$, 80 kelvins, and 10 light years. The mass of each is about a hundred times that of the Sun. Clouds about five times more optically thick than the more prevalent clouds were found to occur about eight times less frequently.

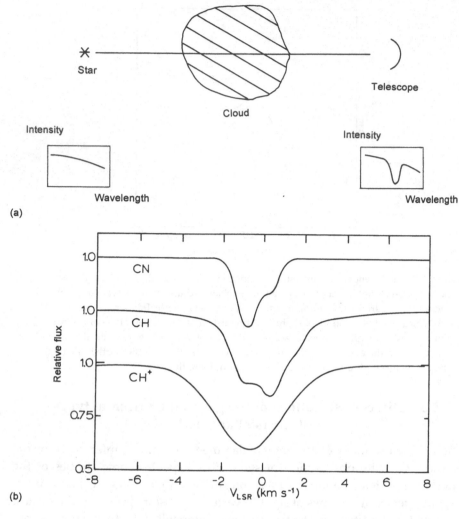

Fig. 5.6. Interstellar absorption lines. (a) If a star were viewed in the absence of intervening gas its spectrum would lack absorption features that the presence of an intervening cloud would cause. (b) Optical absorption caused by interstellar gas towards the bright star, ζ Ophiuchi, demonstrates the presence of interstellar CN, CH, and CH⁺. The lines have been shifted so that we may compare the profiles. The lines are broadened because the emitting molecules have random velocities in the clouds.

Note that to this point we have restricted attention to clouds that can be studied spectroscopically in optical absorption against background stars. We call such clouds *diffuse clouds*. As we shall discuss below, other clouds are too optically thick to be studied with absorption measurements at visual wavelengths. In the late 1930s, it was discovered that *some* (but not all) of the diffuse clouds contain molecules that give rise to features in optical absorption against bright stars (see also Section 1.1). Figure 5.6(b) shows CN, CH, and CH⁺ features detected against a bright background star; they are formed in one of the more optically thick of the diffuse clouds.

The observations of atomic hydrogen at a wavelength of 21 cm also revealed the existence of a phase of the interstellar gas that is warmer, at about 8000 kelvins, and more tenuous, at a number density of about $2 \times 10^5 \, \text{m}^{-3}$, than the interstellar clouds. This 8000 kelvin gas is, as is that in the diffuse clouds, probably heated by the energy carried by electrons ejected from grains following the absorption of starlight. Observations of ultraviolet absorption features of highly stripped ions (e.g. O^{5+}, oxygen that has lost five electrons) and of diffuse soft X-ray emission at about 2.5 nm imply that the interstellar medium also possesses a phase with a temperature of 3×10^5 kelvins to 10^6 kelvins and a number density of the order of $(1-10) \times 10^3 \, \text{m}^{-3}$. It is thought that the hotter gas is energized by the shocks driven into the interstellar medium by supernovae, the explosions that stars more than about ten times as massive as the Sun undergo when they have used most of their nuclear fuel (cf. Chapter 10). Together the 8000 kelvin and roughly million kelvin gases comprise what is called the *intercloud medium*.

The precise fractions of the interstellar volume filled by the hot gas and by the warm gas are hard to determine. Together they occupy all but about 2 per cent of the interstellar volume, and each fills between about 10–90 per cent of it. The clouds take up the remaining small fraction of the space. In any case, the warm gas and hot gas possess pressures that confine, at higher densities, clouds that are at lower temperatures. The confinement of the more optically thin of the diffuse clouds is due only to the pressure of the intercloud medium around them.

The more optically thick diffuse clouds are probably confined primarily by their self-generated gravitational fields, i.e. they are held together by their weight. At earlier times they were almost certainly more tenuous and confined by the pressure of the intercloud medium around them. However, the passage of one of the supernova driven shocks (that energize the hottest interstellar gas) around a cloud will compress it to higher density and after the compressed cloud has cooled, the Jeans criterion for it is more easily satisfied (cf. the discussion of gravitational binding in Section 4.2).

Clouds more optically thick than the diffuse clouds were also known to exist from the obscuration caused by the dust in them, but the understanding of the detailed structures of these so-called *translucent* and *dark* clouds did not develop until the later 1970s and the 1980s when large scale mapping of CO emission in the Galaxy and small scale mapping of CO in some translucent and dark clouds were executed. Each of the translucent and dark clouds is bound together by the gravitational field that it produces. The CO observations showed that the translucent and dark clouds usually possess primarily molecular centres with mostly atomic rims. Some of the translucent and dark clouds have masses comparable to the diffuse clouds and many have been diffuse clouds that were compressed by factors of about 10 by the passages of shocks driven by supernovae.

However, most of the molecular material traced by the CO emission lies in gravitationally self-binding objects called *giant molecular cloud complexes* or simply *giant molecular clouds* which are very likely agglomerations (formed by the collisions between clouds which have a random velocity of about $10\,\mathrm{km\,s^{-1}}$ relative to one another) of clouds that were once diffuse. A giant molecular cloud typically has a mass around a hundred thousand to a million times that of the Sun, very roughly a hundred to a thousand times that of a typical diffuse cloud. The giant molecular clouds contain structures on many scales. Most of the mass is in translucent clumps of hundreds to thousands of solar masses with molecular number densities of the order of 10^9 H_2 molecules per $\mathrm{m^3}$. A giant molecular cloud is typically at least 100 light years across with only about 1 per cent of its volume filled with the massive clumps. Star formation takes place primarily in giant molecular clouds; structures having the various properties associated with objects collapsing from number densities of 10^9–$10^{13}\,\mathrm{m^{-3}}$ have been thoroughly observed in giant molecular clouds as will be described more completely in Chapter 6.

5.4   Molecular hydrogen – the key to interstellar chemistry

We now begin our consideration of chemistry in interstellar space with a rather detailed discussion of H_2. The length of this section is merited because hydrogen is the most abundant element in the Universe, leading one to expect H_2 to be by far the most prevalent molecule in many astronomical environments. Helium atoms are also fairly abundant; there is one helium atom for every ten hydrogen atoms; however, helium is – of course – unreactive. The other main reactive elements of oxygen, carbon, nitrogen, iron, silicon, and sulphur are very much less abundant than hydrogen. Oxygen atoms are present at less than one atom in 1000 hydrogen atoms, carbon at about one in 3000 and nitrogen at about one in 10 000. The other species are even less abundant. Generally, therefore, atoms and molecules are more likely to collide with atoms or molecules of hydrogen than with any other species; thus, hydrogen plays a key role in most astronomical chemistries including that which governs interstellar clouds.

The chemistry depends significantly on the fraction of hydrogen that is molecular. It is a rather inefficient process for atom X to combine with an atom of hydrogen, H, in a single collision, as the collision is over and the atoms have parted before they can lose energy by radiating, allowing them to readjust and exist together as a molecule. By contrast, reactions of atoms of element X with hydrogen molecules may be quite different because when the three atoms are close together they may 'forget' how they were previously arranged and can come out in another form:

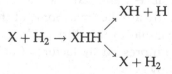

In the first of these reactions a new product has been formed. Excess *chemical* energy will appear as energy of motion of the products, or internal energy in a product molecule.

We might expect that much of the hydrogen is molecular. After all, when two hydrogen atoms combine to form a molecule, energy is released, so H_2 is more stable than H + H. However, although a system generally evolves to seek the lowest energy state, for the chemical conversion from atomic to molecular hydrogen to occur in a reasonable period of time there must be an efficient route by which that conversion can be achieved. For instance, we have seen already that the inefficiency of H_2 formation in the Early Universe (Chapter 4) resulted in the fraction of hydrogen that was molecular being really rather small.

In fact, observations of hydrogen in the interstellar medium indicate that, overall, much of the hydrogen can be molecular. In diffuse clouds, through which starlight readily penetrates, molecular hydrogen is detected on many lines of sight by its absorption at characteristic wavelengths in the ultraviolet. Here, a hot star is used as the background source. The abundance of H_2 can be found from these observations, and varies quite widely. In some cases H_2 molecules are only one millionth as abundant as hydrogen atoms. In others, the amounts of H and H_2 are comparable. In dark clouds, however, we cannot easily detect H_2 by absorption in the ultraviolet since the extinction by dust obscures the background star. We can regard the high extinction as implying the presence of large amounts of gas, but this gas must be almost entirely molecular since very little atomic hydrogen emission at 21 cm is present.

Molecular hydrogen is destroyed in interstellar space by the ultraviolet radiation field from bright stars. Though all interstellar molecules can be photodissociated, the detailed process by which H_2 photodissociation occurs appears to be unique to molecular hydrogen, and was shown schematically in Fig. 4.3. It turns out that for the interstellar radiation field, of the order of 10 per cent of the absorptions lead to the molecule being destroyed in this way. A single H_2 molecule exposed to an average ultraviolet radiation field therefore has a lifetime expectation of less than a thousand years: not bad by terrestrial standards but in interstellar terms only about 1 A-minute.

It is now possible to ask: will the mechanisms that made H_2 in the Early Universe also make enough H_2 in the interstellar medium? These mechanisms involved electrons and protons as catalysts (see Section 4.1). In the interstellar clouds, electrons and protons are relatively minor constituents with abundances relative to hydrogen of less than about one in ten thousand. These routes for forming H_2 are really rather inefficient in interstellar environments. In fact, using these mechanisms, we cannot explain an abundance of H_2 greater than about one part in a billion of the H atom abundance. This is one thousand times smaller than the typical lowest H_2 abundances. So we have two problems: firstly, we need to identify a new more efficient mechanism that can produce H_2 at least a thousand times faster than the electron and proton catalysis mechanisms; and, secondly, we need to understand how it is that in some regions almost all the hydrogen (more than 99 per cent) is molecular.

The new mechanism, not available in the Early Universe, is H_2 formation on dust. The idea is straightforward. An H atom collides with a dust grain and is weakly bound to the surface; a second H atom arriving at the grain finds the first, they combine to form a molecule, and the energy which must be released in the process is absorbed by the grain. Recombination of atoms at surfaces is a common phenomenon in nature; in the interstellar medium it needs to be fairly efficient to account for the observed H_2 abundances at about 10^{-6} of the total hydrogen abundance. Since we know something about the grains we can work out how efficient the process must be: it turns out that nearly every H atom arriving at a grain must leave as part of a molecule of H_2.

So much for the formation mechanism: no faster mechanism has ever been proposed. How is it, then, that the relative abundance of the ratio H_2:H can change from about 10^{-6} to about 10^2? It must have something to do with the destruction process. Obviously, the presence of dust helps: it shields the region from the destroying ultraviolet radiation. But this alone is not enough to account for the enormous range in H_2:H. In fact, the molecular hydrogen solves its own problems: it shields itself! It can do this because the destruction mechanism (as shown in Fig. 4.3) starts with excitation from the ground electronic state to an upper electronic state by the absorption of radiation. This excitation occurs in a number of very narrow bands, around one hundredth of a nanometre wide. Radiation outside these very narrow bands does not destroy the molecules. Where a molecule at the edge of a cloud has been destroyed, the radiation in this narrow band is reduced in intensity, and many such dissociations rapidly reduce the intensity in these narrow bands to a point at which destruction of H_2 becomes slower than its formation. Further into the cloud, the hydrogen is largely molecular.

Radiation emitted from H_2 is observed from shocked gas in star forming regions (see Section 6.6). Otherwise, H_2 in the interstellar clouds has normally been observed primarily in ultraviolet absorption arising in diffuse clouds with bright background stars, as we mentioned earlier in this section. Those observations permit the construction of detailed models of diffuse clouds. Absorption from the populations in up to eight of the lowest rotational levels has been detected. The populations of the highest of those levels give information about the radiation field incident on a cloud since those levels are populated following ultraviolet absorption and a subsequent radiative cascade to and through excited vibrational and rotational levels in the ground electronic state of H_2. The relative abundances of H_2 and H depend on the radiation field (the properties of which have already been constrained from the populations of the higher observed rotational levels) and the cloud density, and consequently are used to derive the density. The populations of the lower rotational levels are determined to a large extent by the rates at which collisions induce excitation; since the efficiency of collisional excitation is determined in part by the temperature, the observed populations of the more lowly excited rotational levels of H_2 yield information about a cloud's thermal structure. Thus, H_2 observations are used to determine the detailed properties of diffuse clouds containing molecules.

H_2 in cool translucent and dark clouds has not been observable. Thus, models of those clouds' physical properties are based on other molecules;

often CO is used since emissions from several of its levels are observable and provide diagnostics. Other species are also studied to unravel the properties of translucent and dark clouds.

5.5 Dark clouds: chemical factories in the interstellar medium

The dark clouds, those through which the intensity of optical radiation is reduced by a factor of more than 100, are now known to contain a great variety of molecular species (see Table 1.1). Different regions have rather different chemistries, and we should be able to use this information to infer the different histories and conditions in them. The molecular cloud in Orion is a region where many of the molecules listed in Table 1.1 have been detected. It is a region of great activity – a centre of astronomical industry; this molecular cloud is close to the bright Orion Nebula pictured in Fig. 1.1. Here, massive hot stars are being forged from interstellar gas. (See Section 7.6 for more on star formation in Orion.) The number density in this molecular cloud is about $10^{10}\,\mathrm{m}^{-3}$, or 10 000 times denser than the interstellar average. The cloud is very cold, with temperatures of 10–30 kelvins. It contains so much dust that it is optically opaque to visible and ultraviolet light. Undoubtedly, this high extinction contributes to the chemical variety, for the molecules in these dark clouds are shielded from the powerful and destructive interstellar radiation field.

In this section we shall describe how chemistry involving hydrogen, carbon, nitrogen and oxygen is initiated in dark clouds. We shall assume that the hydrogen is nearly all molecular, and that oxygen, carbon, and nitrogen are initially atomic and neutral. We shall ignore any radiation field on the assumption that the dust excludes it totally. In fact, the only significant and useful energy source penetrating the interior of these regions is particle radiation, the cosmic rays, rather than electromagnetic radiation. Cosmic rays are particles travelling at speeds above about a tenth of that of light. Of all cosmic rays, cosmic ray protons have the most important effects on the gas in clouds. Their main effect on the interstellar gas is to cause a slow rate of ionization as they collide with H_2 molecules and eject electrons to form H_2^+ 97 per cent of the time and H^+ only 3 per cent of the time. The ion H_2^+, arising in nearly all the cosmic ray ionizations of H_2, almost always reacts with H_2, the most abundant species, to give a new ion, H_3^+. The system is illustrated schematically in Fig. 5.7.

At low temperatures atoms of oxygen do not react with molecular hydrogen: neither do atoms of carbon, nor those of nitrogen! The obvious entry routes into the chemistry are forbidden. Nature is clearly more subtle! Reactions of the H_3^+ ion with atoms of oxygen and carbon provide the necessary route.

The ion H_3^+ is reactive and can donate a proton to many species. For example, it can give a proton to an oxygen atom to form OH^+ and leave one molecule of H_2 behind. The OH^+ ions are also reactive, and can extract a hydrogen atom from a hydrogen molecule to form H_2O^+ which reacts with H_2 to form H_3O^+.

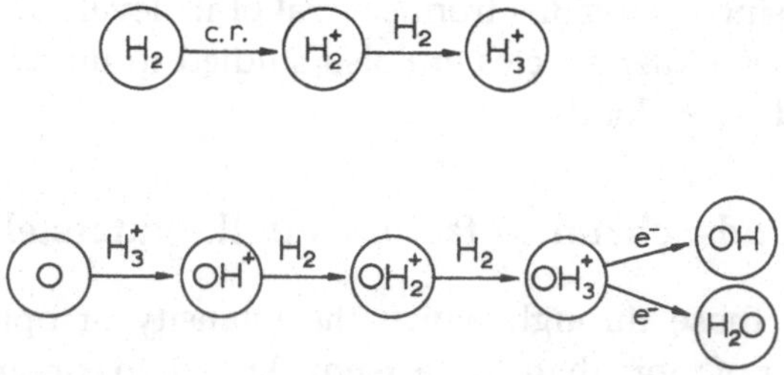

Fig. 5.7. A schematic diagram of the backbone of the H_3^+ and oxygen chemistries in dark clouds.

Further reactions of this type do not occur, however, as the oxygen ion cannot form any further chemical bonds; its valencies are fully saturated in H_3O^+. The ion, H_3O^+, is likely to be neutralized in collisions with electrons, since the electrostatic interaction between ions and electrons is very strong, and the molecules OH and H_2O are both formed with H_2O being rather more likely. These are two of the observed species. This sequence of reactions is illustrated in Fig. 5.7, part of which bears some similarity to Fig. 3.8.

The backbone chemistry of carbon in dense clouds is in many ways similar to that of oxygen, with H_3^+ playing a significant role in initiating it. One can easily see how CH_3^+, CH, and CH_2 will form in analogy to H_3O^+, OH, and H_2O. Once these backbone chemistries of oxygen and carbon are established, reactions between their products can proceed. For instance, a significant route to the formation of one of the most important interstellar molecules is

$$CH_3^+ + O \rightarrow HCO^+ + H_2,$$

$$HCO^+ + e^- \rightarrow CO + H.$$

CO is one of several relatively nonreactive species that are easily formed. If there were no destruction mechanism it would eventually contain all of the gas phase carbon. In fact, CO is destroyed by helium ions. Cosmic rays ionize helium to make He^+; these ions may then react with CO to produce C^+. These C^+ ions can then react with H_2 to initiate the backbone carbon chemistry again, with OH or H_2O to form CO again, or with other trace species to trigger other branches of chemistry.

The simple idea of a 'backbone' chemistry initiated by neutral atom reactions with H_3^+ and followed by a sequence of ion–H_2 reactions is not applicable to all elements. At dense cloud temperatures, H_3^+ simply does not react with all neutral atomic species. For instance, the nitrogen chemistry is initiated in dark clouds by atomic nitrogen reacting with OH and CH to form NO and CN.

Section 2.1 contained a mention of chemical timescales and of how they can be shorter than or longer than the timescales on which various physical properties in an astronomical source change. We shall return to these sorts of considerations in Chapter 6 which deals with star formation. However, we note here that one of the important chemical timescales in a dark cloud is simply the length of time required to ionize an amount of hydrogen comparable to the amount of oxygen and carbon in the gas phase. If we assume that some of the oxygen and carbon is unavailable, being locked in solid form in the dust, so that there are roughly ten thousand hydrogen molecules for each atom of gas phase carbon and oxygen, this length of time is one ten thousandth of the timescale required for cosmic rays to induce each hydrogen molecule to be ionized once. This cosmic ray induced ionization timescale is roughly ten billion years; one ten thousandth of that is about a million years. Therefore, if dark clouds exist for a million years or so, very substantial chemical processing will occur within them. This processing is driven by the ionization caused by cosmic rays.

5.6 Diffuse clouds: chemistry by starlight

Diffuse clouds are pervaded by starlight: although some extinction is caused by the dust so that the ultraviolet radiation intensities are reduced by factors of around 10, the destructive effects of the radiation remain powerful. We are familiar with this effect in many ways. For example, sunlight bleaches the curtains in our houses by altering the structure of the dye molecules in the fabric. Excessive sunlight destroys the cells that make up human skin. It is therefore not surprising that the chemistry of diffuse clouds is neither as rich in variety, nor as abundant, as in dark clouds. In addition to CH, CN and CH^+ (see Chapter 1) H_2, CO, and OH are commonly detected.

Is the chemistry different in character from that in dark clouds? Apart from the rapid destruction of molecules by photodissociation (the average life of a molecule in a diffuse cloud is about 300 years, or less than half an A-minute) much of the description of dark cloud chemistry applies also to diffuse clouds. However, the abundance of electrons is now much higher because the radiation field ionizes carbon atoms (though not those of oxygen and nitrogen; radiation that is capable of ionizing oxygen and nitrogen also ionizes hydrogen, so that oxygen and nitrogen are shielded by the much more abundant hydrogen). Although C^+ ions cannot react rapidly with H_2 except at temperatures over about 1000 kelvins, in about one in a million collisions the complex can stabilize by emitting radiation

$$C^+ + H_2 \rightarrow (CH_2^+)^* \rightarrow CH_2^+ + \text{radiation}$$

and the resulting CH_2^+ ion reacts with H_2 to form CH_3^+ which then, as in dark clouds, is neutralized by the capture of an electron to form sometimes CH and sometimes CH_2.

Since oxygen atoms are not photoionized by the radiation in diffuse clouds, and since the H_3^+ abundance is low (because the electron abundance is high) we need another entry into oxygen chemistry. This is provided through the O^+ ions created, indirectly, by cosmic rays which ionize hydrogen to produce H^+. It takes nearly the same amount of energy to remove an electron from a hydrogen atom as from an oxygen atom. Therefore, if an oxygen atom and a hydrogen ion collide at temperatures around 100 kelvins or more, then it is quite likely that the charge will transfer

$$H^+ + O \rightarrow O^+ + H.$$

This is a ready source of O^+ which reacts with H_2 to form OH^+ and to feed the chain of reactions that produces OH and H_2O. The chain of reactions is driven by cosmic ray ionization of H atoms; each proton produced leads to the formation of OH or H_2O. However, these products do not survive long in the radiation field, so they too are not very abundant. Less than one hundredth of 1 per cent of the available oxygen is tied up in OH and H_2O.

Clearly, all of the reaction schemes involve molecular hydrogen to a great extent. If they can, ions will react successively with H_2 and extract a hydrogen atom each time. When no more hydrogen can be added, the resulting ions will recombine with electrons to give neutral hydrides which can react with other species containing heavy atoms to make molecules with several heavy atoms. This is very similar to the dark cloud chemistry, except that the radiation field is a major source of ionization and is very efficient at destroying molecules. We expect, therefore, that a successful diffuse cloud chemistry is possible only where much of the hydrogen is molecular. This is generally true: along lines of sight where molecules are detected in diffuse clouds the number of hydrogen molecules is approximately equal to the number of hydrogen atoms and in some parts of the clouds almost all hydrogen is molecular. In those parts of the cloud an ion has a good chance of colliding with H_2. Clouds in which H_2 is a minor species do not in general show many other molecules. The molecular ion CH^+ is, however, sometimes an exception.

5.7   Dust: the interstellar catalytic converter

Without dust, molecular hydrogen could not be abundant in the Galaxy. Interstellar chemistry would then be rudimentary, and interstellar clouds could not evolve as they do to form stars of a wide range of masses, and planets. We may ask: do dust grains have other effects, in addition to H_2 formation, on interstellar chemistry?

One important effect, potentially dominating, is that in dark clouds molecules other than H_2 tend to freeze out on dust grain surfaces. There is nothing strange in this: in the laboratory it is difficult to keep any surface truly clean. Water molecules in the air tend to coat any surface. In space, the dust usually is very cold indeed, typically about 10 kelvins. Molecules tend to stick when they collide with grain surfaces, and icy mantles accumulate on them. Even atoms such as oxygen, carbon, and nitrogen will stick, and are probably converted by hydrogen addition to water, methane, and ammonia. Such mantles are detected by spectroscopy in the infrared. For example, water ice absorbs in the laboratory quite strongly at a wavelength of 3 micrometres. A similar absorption feature is often observed in the interstellar medium when the infrared radiation from a cool star passes through a dark cloud. From such observations we can deduce that in many dark clouds

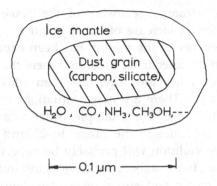

Fig. 5.8. The core–mantle structure of a dust grain in a dark cloud. The icy mantles do
not build up in diffuse clouds where photons may destroy them.

much more H_2O is in ice than in the gas. Another molecule detected to be
present in the icy mantles is CO. When it is frozen in a matrix it absorbs
infrared radiation at a wavelength near 4.7 micrometres. Figure 5.8 indicates
schematically the dust grain and mantle.

Of course, if molecules such as H_2O and CO freeze out unrestrictedly, then
eventually nearly all these 'heavy' molecules will be in the solid. This is not
the case, so some process must be returning the ices to the gas phase. Perhaps
some process induced by star formation does this intermittently, or perhaps
there is a continual evaporation of molecules from the mantle.

Nevertheless, substantial amounts of oxygen, carbon, and nitrogen are, in
various forms, locked up in mantles on grains. In dark clouds, perhaps
around half of the atoms of these elements are typically in solid form. Laboratory work has shown that if such ices are exposed to ultraviolet light – as may
happen if a new star is formed inside a molecular cloud – then new chemical
species are formed. These experiments show, for example, that methanol,
CH_3OH, forms efficiently from a mixture containing CO and H_2O. So a whole
new range of products becomes possible as a result of solid state chemistry,
when icy mantles on dust grains are energized by stellar radiation. Where
this happens, the grains will tend to become warmer, and the icy mantles
may be evaporated, populating the gas with a greater variety of molecules.
The high abundance of ammonia, NH_3, detected in some dense regions of
dark clouds, probably arises as a result of hydrogenation of nitrogen atoms.

Icy mantles are not detected in diffuse clouds, so the solid state chemistry
energized by ultraviolet radiation does not seem to be operating there. It
remains a possibility that atoms are converted to their hydrides on collision
with dust. The recent detection of NH in a diffuse cloud seems to require the
presence of chemically active grains.

5.8   Interstellar shocks: overcoming the chemical hump

Shocks occur whenever a disturbance is driven through a fluid at a speed
greater than the sound speed. As we described in Section 4.3, the air ahead

of a supersonic plane cannot be aware of the plane's approach and the consequence is an abrupt increase of density and temperature, i.e. a shock wave. Some of the energy of the motion has been expended in accelerating the postshock gas. In the interstellar medium where the gas is largely neutral and temperatures are fairly low, say, less than 100 kelvins, the speed of sound is about $1 \, km \, s^{-1}$. There are many situations where velocities many times larger than this arise. For example, interstellar clouds often have apparently random velocities in the range $10–20 \, km \, s^{-1}$. If they bump into one another, then the collision will probably be supersonic and shocks will occur. Flows of gas from stars also have substantial velocities. Stellar explosions such as novae and supernovae also generate very substantial motions of gas which are always accompanied by shocks.

The temperature rise behind a shock is abrupt and can be large. For a shock velocity around $10 \, km \, s^{-1}$ we might expect a postshock temperature around a thousand kelvins, quite warm by terrestrial standards. Faster shocks with speeds of several tens of kilometres per second or more will create even higher temperatures, sufficient to dissociate molecules.

Many chemical reactions require a temperature rise to start them going: a lump of coal surrounded by air does not spontaneously ignite; petrol vapour in the cylinder of an engine does not of its own volition explode and drive down the piston; coal needs to be heated in air; petrol vapour needs the ignition of the spark. Although the products of combustion represent a lower energy state than the reactants, there is an energy barrier between reactants and products. This barrier can only be overcome by the addition of some energy to the reactants.

Until now in our discussion of chemistry in interstellar space we have excluded many reactions of this type, i.e. with barriers. In particular, we have stated that reactions of the atoms O, C, and N with H_2 are suppressed at the low temperatures, less than about 100 kelvins, generally found in interstellar clouds. Although the reaction of O with H_2 is *exothermic*, i.e. energy is released in the reaction, it is impeded by an activation energy barrier equivalent to about 1000 kelvins. In an interstellar shock of modest velocity such temperatures are readily achieved. Then we may expect H_2O to be formed very rapidly in reactions in which O and OH extract H atoms from H_2, so that all of neutral atomic oxygen (rather than just the very minor components O^+ and OH^+) feeds the production of H_2O. We therefore expect a conversion of oxygen to H_2O to occur in shocks of modest speed ($10 \, km \, s^{-1}$, say). The cases of C, C^+, and N are different: these reactions are *endothermic* – energy must be added if the reactants are to be able to reach the product state. But the energy deficit is easily made up at temperatures achieved in shocks of modest velocities. Therefore, we expect atomic nitrogen to be converted to ammonia, carbon atoms to make simple hydrocarbons, and carbon ions to make hydrocarbon ions in such regions.

If shock speeds and densities are high enough, collisions will dissociate H_2 to form H atoms. Hot H atoms are quite destructive. If much of the hydrogen is dissociated then the chemistry is reversed, i.e. a reaction such as

$$N + H_2 \rightarrow NH + H$$

is opposed by the reverse reaction

$$NH + H \rightarrow N + H_2.$$

Thus, for dense regions there is a critical shock speed above which chemistry is suppressed entirely. This is several tens of kilometres per second. In any case, shocks with speeds much above $100\,\mathrm{km\,s^{-1}}$ will dissociate and ionize most hydrogen in a medium of even quite low density.

5.9 Chemical influences on cloud temperatures

So far in this chapter we have considered the nature of interstellar clouds and interstellar chemistry. We now begin to show how chemistry controls the *evolution* of the clouds.

Section 3.10 contains an introduction to the basic mechanism of radiative cooling initiated by the population of molecules in collisionally excited rotational and vibrational levels. This general mechanism dominates cooling in all molecular clouds, but which molecular species radiates the most energy depends on cloud conditions. If the lowest excited level of a molecule has an energy that is much more than a few times the average kinetic energy of a molecule in a gas (which is exactly what temperature measures) then that molecule usually does not play a role in cooling.

There are four species of molecule that are important coolants in at least some interstellar cloud environments. They are CO, OH, H_2O, and H_2. Primarily because it is the heaviest, CO has the least energetic first excited rotational level state of all of these molecules. Its first excited level lies at an energy of only 7.6×10^{-23} joule; for comparison a particle in a gas with a temperature of only about 3.6 kelvins on average possesses 7.6×10^{-23} joules of kinetic energy. The lowest rotational level of OH has an excitation energy about 20 times higher and the lowest energy level that can be excited in an H_2–H_2 collision is at an energy about 100 times higher than the lowest level of CO. Thus, many collisions of H_2 molecules with CO molecules excite the CO even when the temperature is as low as a few kelvins, whereas excitation of OH and H_2 at appreciable rates occurs only when the temperature rises to several tens of kelvins and around a hundred kelvins, respectively. The lowest excited rotational level of the H_2O molecule is at an energy lower than that of OH, but higher than that of CO, and water, if abundant, may contribute as a coolant at temperatures as low as about ten kelvins. However, it is probably never a more important coolant than CO in such cool environments, partly because it is usually considerably less abundant in cold gas.

In the discussion above, we have already established that CO will be the most important coolant in cold (less than several tens of kelvins) gas, largely because it has rotational levels that are the most closely spaced of the most abundant molecules. In general, the more collisions there are between H_2 and CO the more cooling occurs. However, two effects can act to reduce the CO cooling rate per unit volume far below the value that would obtain if every excitation-inducing collision were to result in the emission and escape of a photon. Firstly, if the number density of H_2 is too high (about $3 \times 10^8 \, m^{-3}$ or higher), collisions often depopulate the excited levels of CO before they radiate; when collisionally induced deexcitation occurs the energy simply becomes thermal in nature again. Secondly, if a great deal of CO exists throughout the cloud, radiation emitted by the decay of excited CO in the cloud can be absorbed by unexcited CO also in the cloud; i.e. the cooling radiation is *trapped* in the cloud so the cooling rate is reduced. Column densities of CO at which trapping effects become important do not obtain in most diffuse clouds, but they are common in translucent and dark clouds.

As the temperature increases, coolants other than CO can, in principle, become important, because enough H_2 molecules possess the energies required to induce collisionally the population of excited levels of those other species. However, CO is not negligible as a coolant until temperatures well over several hundred kelvins are reached. As the temperature increases from 10 kelvins many more excited levels of CO can and do become populated and radiate. H_2, the only species much more abundant than CO, does not start to become an important coolant until temperatures over several hundred kelvins are reached. This is because radiative transitions between the rotational and vibrational levels of H_2 are much slower than those of other coolant molecules.

Eventually, at sufficiently elevated temperatures, H_2, OH, and H_2O become the dominant coolants. As mentioned above, H_2 radiates slowly. Thus, OH and H_2O (which may be produced by the sort of shock chemistry described in Section 5.8) compete with H_2 as coolants. OH is usually a more important coolant than H_2O in diffuse shocked regions where the temperatures are typically about a thousand kelvins. However, because radiation is not so important in determining the chemistry of dark clouds, H_2O is more abundant and a more important coolant than OH in shocked regions in dark clouds. OH and H_2O cooling rates can be affected by radiation trapping, as in the case of CO cooling. In diffuse cloud shocks, radiative trapping is unimportant and the OH cooling rate per unit volume is what would be expected if every excitation were to lead to the emission of radiation that escapes. However, the H_2O cooling rate per unit volume in dense cloud shocks is reduced by the effect of radiative trapping.

At high enough temperatures, radiative losses from excited vibrational levels of H_2 become important. H_2 therefore is the dominant coolant in shocks with temperatures of a couple of thousand kelvins, as long as the density is not too high. In interstellar clouds radiative trapping does not affect H_2 cooling, but at number densities much above 10^{11}–$10^{12} \, m^{-3}$ (depending on the temperature in the shocked gas) collisions start to compete with radiative processes in the depopulation of the H_2 excited vibrational levels. At higher densities H_2O is usually the most important coolant.

In Section 5.2, we described how the ejection of energetic electrons from grains as they absorb stellar light acts as a heating process in diffuse clouds. This heating balances the cooling in molecular regions of diffuse and translucent clouds to maintain temperatures in the range 20–30 kelvins. Some molecular gas in diffuse clouds is warmer. We have described shocks as one way of heating gas, but other mechanisms probably operate as well.

Interstellar radiation does not easily penetrate to the interiors of dark clouds, so the heating mechanism there is not the same as for diffuse clouds. An alternative, less efficient mechanism is the cosmic ray induced ionization heating, described in Section 3.10. It seems unlikely that it is responsible by itself for maintaining dark clouds at the observed temperatures of 10–30 kelvins. (Shocked dark regions are, of course, much hotter.) Some dynamical heating mechanism may be balancing the CO cooling that dominates in the low-temperature dark regions. One is possibly frictional heating driven by ambipolar diffusion, a magnetohydrodynamic process described towards the end of the next section.

5.10   Magnetic retardation of cloud collapse in regions of stellar birth

As described in Section 5.3, the most diffuse interstellar clouds are confined by the pressure of the warm and hot gas around them, while the more optically thick interstellar clouds are dense enough that their gravity binds them together. So far, we have mentioned thermal pressure as the only type of pressure playing a role in cloud support or confinement (see the discussion in Section 4.2 of the Jeans mass and Section 5.3). However, the interstellar gas contains a *magnetic field*. We know from the way in which a magnet can be used to pick up metal filings against the force of the Earth's gravity that magnetic fields can exert force. Pressure arises when a force acts over a surface area, and the interstellar magnetic field possesses a corresponding 'magnetic pressure' which is always at least comparable to the thermal pressure and which sometimes greatly exceeds it. Hence, we must modify our original picture of cloud confinement and support to include magnetic effects. We shall see that, as for the gas pressure which is affected by the presence of molecular coolants, the effectiveness of the 'magnetic pressure' is controlled by the chemistry. The effectiveness of the magnetic pressure in many clouds determines whether collapse within them will lead to regions of stellar birth.

Two possible static, pressure-confined cloud configurations are illustrated in Fig. 5.9. Figure 5.9(a) shows a cloud for which gravity is unimportant and in which the magnetic field is the same as it is in the external medium. The strength of a field is proportional to the number of field lines passing through a surface of a fixed area; i.e. where the field lines are close together, the field strength is high, and vice versa. Hence, the constancy of the field is represented by the equal spacing between field lines. The force due to the magnetic field is nonzero only if the magnetic field strength varies; the force always is directed perpendicular to the

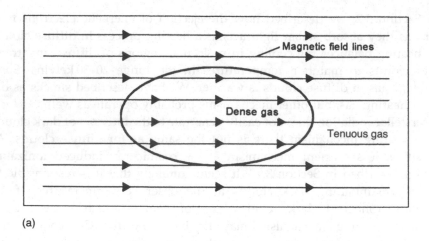

(a)

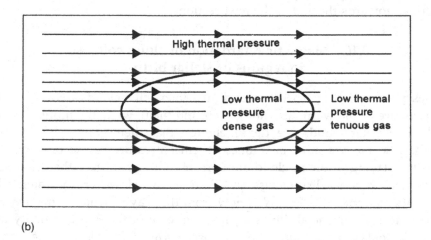

(b)

Fig. 5.9. Magnetized pressure-confined clouds. (a) The thermal pressure is constant, and the magnetic pressure is constant. (b) The sum of the magnetic and thermal pressures is constant, but the magnetic pressure varies as does the thermal pressure.

magnetic field lines. In the configuration pictured in Fig. 5.9(a), the magnetic force is zero because the magnetic field is constant. Hence, the thermal pressure of the cloud and the thermal pressure of the surrounding gas must be equal to one another, since we have assumed that gravity is unimportant, i.e. the cloud is much less massive than the Jeans mass.

Figure 5.9(b) shows a configuration in which the magnetic field is constant above and below the cloud but is stronger inside the cloud. As depicted, the field strength is constant in the cloud and does not vary with position along the magnetic field direction. The magnetic force is always directed perpendicular to the field lines. It is zero above and below the cloud because the field strength is constant. It is also zero inside the cloud, but it is not zero over the top and bottom surfaces of the cloud. At those surfaces, the field lines

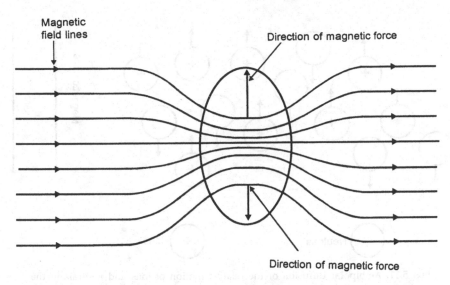

Fig. 5.10. A gravitationally bound, magnetized cloud. The magnetic field creates a force that helps to support the cloud against its own weight.

change from being close-packed to widely spaced. This means that the field strength changes over those parts of the cloud boundary, so there is an outward magnetic force, balanced by a change in the thermal pressure. The increase in the thermal pressure occurs only as the field lines are crossed. The thermal pressure along an individual field line does not change since no magnetic force exists parallel to a field line.

We now consider magnetized clouds for which gravity is important. The magnetic field can help support a cloud against gravity, so that the *magnetic Jeans mass* (i.e. the minimum mass of a cloud with a given density, temperature, and magnetic field strength that will collapse due to its gravity) is greater than the nonmagnetic Jeans mass given in Section 4.2. Even if the magnetic field is very large the ratio of the magnetic and nonmagnetic Jeans masses is limited because gravitationally driven collapse along the magnetic field lines is hindered only by the thermal pressure, which is also important in the nonmagnetic case. If the magnetic field is very strong, gravitationally driven collapse will initially occur along the field lines. Eventually the density will become so high that the gravity will be strong enough to drive collapse perpendicular to the field lines. A nearly static cloud configuration is often the result of such an evolution. A nearly static cloud supported against gravity by thermal pressure and magnetic fields is drawn schematically in Fig. 5.10.

We refer to the cloud being 'nearly static' because, in fact, the magnetic field influences the distribution of the neutral material only when small relative velocities between the neutrals and the charged particles exist. Figure 5.11 shows schematically the directions of the magnetic force and the gravitational force in the upper half of the cloud depicted in Fig. 5.10. The magnetic force acts only on the charged particles and tends to push them outwards. (The gravitational force on the charged particles is much weaker than the magnetic

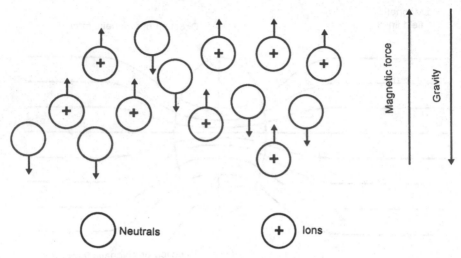

Fig. 5.11. Ambipolar diffusion or the relative motion of ions and neutrals in the presence of a gravitational field and a magnetic field.

force on them.) The magnetic force does not act directly on the neutral gas, and gravity pulls them inwards. The outward motion of the ions and electrons and the inward motion of the neutrals give rise to friction between the charged particles and the neutrals; this friction acts to reduce the outward velocity of the charged particles and the inward velocity of the neutrals. Because of this friction the presence of the magnetic field leads to a retardation of collapse; it does not lead to the existence of a strictly static cloud configuration. The collapse of a weakly ionized cloud can remain magnetically retarded for a time that is at most the time required for the neutrals to drift, in response to the gravity, through the ions and electrons and other charged particles. This drift of neutrals relative to charged particles is called *ambipolar diffusion* and occurs on a timescale of 400 000 years in a cloud in which the collapse is magnetically retarded and in which the fractional ionization (the ratio of the number density of all ions to that of hydrogen nuclei) is 10^{-8}. The ambipolar diffusion timescale increases in proportion to the fractional ionization and is 4 million years for a fractional ionization of 10^{-7}. The ambipolar diffusion time is, therefore, the maximum timescale for which a cloud can remain nearly static. Then collapse leads to the formation of the regions of star formation described in Chapter 6.

5.11   Ionization in clouds – the chemical control of magnetically retarded collapse

As described in the previous section, magnetic retardation of cloud collapse is, therefore, effective for a timescale that depends on the fractional ionization in the cloud. What controls the ionization in a cloud? We shall see now that it is the chemistry.

For regions deep enough into a cloud that the hydrogen is mostly H_2 but not so far in that the CO contains most of the carbon, the dominant ion is

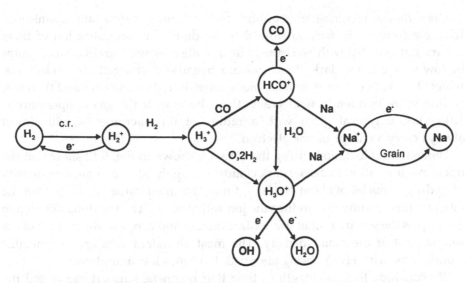

Fig. 5.12. Chemistry controlling the ionization in a dark cloud. Sodium has been
chosen as the representative heavy metal.

simply C^+. Radiation does not ionize H and O here, because the radiation that
can ionize them is absorbed by hydrogen near the stars which are the radiation
sources. Carbon is the next most abundant element after hydrogen, helium, and
oxygen and can be ionized by radiation of longer wavelength than is required
to ionize hydrogen. Hence, at moderate depths into a cloud, the photoionization of C, producing C^+, is the major source of ionization (see Section 5.8).

At greater depths into a cloud, those to which very few photons impinging
on the cloud can reach because they are absorbed by dust, the dominant
source of ionization is cosmic rays (see Section 5.5). The cosmic rays ionize
H_2 to form H_2^+ which then reacts with H_2 to form H_3^+ which then combines
with CO, the most abundant species after H_2, to form HCO^+. This sequence
is illustrated in Fig. 5.12.

The ion HCO^+ is often the most abundant molecular ion in dark regions. As
indicated in Fig. 5.12, it can be removed in three main ways. It dissociatively
recombines with electrons to give CO, it reacts with H_2O to form H_3O^+, and
it gives up its positive charge to neutral metallic atoms such as sodium and
magnesium, to form Na^+ and Mg^+. The reactions with the metallic atoms are
extremely important. The ions Na^+ and Mg^+ combine only very slowly
indeed with electrons, perhaps a factor of 10^5 times more slowly than HCO^+
combines with electrons. Since they are also unreactive at low temperatures,
once these metal ions have been created they exist for a long time. This
means that when HCO^+ can give its charge to metal atoms the level of ionization in the cloud is greater than if sodium and magnesium were not present.
The ion H_3O^+ also exchanges its charge with sodium and magnesium in the
same way as HCO^+. Inclusion of H_3O^+ formation in the network makes
some difference since it dissociatively recombines with electrons even more
rapidly than HCO^+, although the difference is small.

The radiative recombination of Mg^+ and Na^+ and other metallic atomic ions with electrons is, in fact, so slow that the dominant neutralization of them occurs not directly with electrons, but in collisions with grains. Most grains in low-temperature dark cloud gas are negatively charged since electrons, which are much lighter than ions, move roughly 100 times faster and therefore collide more frequently with grains than the ions. If the gas temperature is 10 kelvins, a typical grain with a radius of 0.1 micrometres will almost always carry a charge of one electron.

Theoretical calculations using the network shown in Fig. 5.12 show that the fractional ionization in dense, dark clouds is roughly 10^{-8} for a number density of hydrogen nuclei of about $10^{11}\,m^{-3}$, for an assumed value of $10^{-17}s^{-1}$ for the rate at which cosmic rays induce the ionization of H_2. The fractional ionization goes up as the square root of this ionization rate and drops as the inverse of the square root of the cloud density. The most abundant ions are the metallic atomic ions, with HCO^+ being about 10–100 times less abundant.

We conclude that the length of time that magnetic support can retard the collapse of a gravitationally bound cloud depends on the fractional ionization which, in turn, depends on the cosmic ray induced ionization rate. The next step in understanding the chemical control of the magnetic retardation of cloud collapse to produce star forming regions is to infer the ionization state induced by cosmic rays on the gas in interstellar clouds.

5.12 The inference of the cosmic ray induced ionization rate in diffuse clouds

Although the cosmic ray ionization rate is not well known in dark clouds, we can determine it quite well for diffuse clouds. We do this from observations of molecules: once again, interstellar chemistry proves its utility. In diffuse clouds, cosmic rays cause the ionization of species that require more energy to be ionized than can be provided by the starlight reaching the diffuse clouds. The cosmic ray induced ionization drives chemistry; the products of that chemistry are, therefore, a measure of the cosmic ray ionization rate.

Starlight penetrates diffuse clouds with little absorption, and ionizes carbon, but not oxygen and hydrogen. Since neutral atomic oxygen does not react with hydrogen at low temperatures, the entry into oxygen chemistry (see Section 5.6) is through reactions of O with H^+ and H_3^+ which are produced as a result of cosmic rays interacting with neutral H and H_2. The oxygen chemistry forms OH at a rate that depends directly on the rate at which cosmic rays induce ionization. The dominant OH removal mechanism is photodissociation, the rate of which is known because the radiation field is determined using the H_2 and H data (see Section 5.4). Given the cloud density and temperature structures (as can be inferred from H_2 observations as well; see Section 5.4), the photodissociation rate of OH, and the measured OH abundance, then the cosmic ray ionization rate can be calculated for a diffuse cloud.

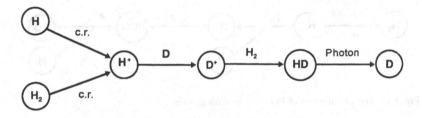

Fig. 5.13. HD formation and removal in diffuse clouds.

We can also use hydrogen deuteride, HD, in the inference of the cosmic ray ionization rate in diffuse clouds. Hydrogen deuteride is chemically similar to H_2, but has one of the H atoms replaced by a deuterium atom, D, which is twice as massive as an H atom but also contains only one proton. The formation and destruction mechanisms of HD are shown in Fig. 5.13. Cosmic rays ionize H and H_2 to give H^+ which transfers its charge to D to form D^+. The deuterium ion then reacts with H_2 giving HD. In diffuse clouds, HD is removed primarily by photodissociation, the rate of which is known because the radiation field intensity is determined from the H and H_2 absorption data. Given the cosmic abundance of deuterium, the cloud density and temperature structures, the HD photodissociation rate, and the measured HD abundance the cosmic ray ionization rate can be inferred for a diffuse cloud.

Cosmic ray induced ionization rates inferred from diffuse cloud data for OH and HD are generally around $10^{-17} \, s^{-1}$, i.e. a given hydrogen molecule or atom in a cloud is likely to be ionized once by a passing cosmic ray in roughly the period of time that the Earth has existed. This seems an incredibly slow rate of ionization. However, for cosmic rays to drive interesting chemistry they must only ionize (at most!) about as many hydrogen molecules as there are oxygen atoms in the gas phase; this can be done on less than a thousandth of the Earth's lifetime.

In one interesting application of these methods of deducing the cosmic ray induced ionization rate, the rates in different parts of the shell of an old (about 1 million years) remnant of a supernova were inferred. (A supernova remnant is the bubble blown in the interstellar gas by a stellar explosion. Typically, a remnant expands to a radius of several hundred light years and loses much of its energy on the timescale of a million years by radiating in the X-ray and extreme ultraviolet regimes.) From the variation of the rate with shell position the conclusion was drawn that many of the ionizing cosmic rays have speeds around 4 per cent that of light. This conclusion has been used as a clue to the way in which the magnetic irregularities that influence the energies and propagation of cosmic rays in the Galaxy are produced and evolve in supernova remnants.

We shall assume for the moment that the value of the cosmic ray ionization rate obtained for diffuse clouds applies also to dark clouds. With a cosmic ray induced ionization rate of $10^{-17} \, s^{-1}$ the discussion of the previous section tells

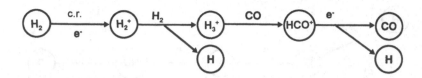

Fig. 5.14. The production of H atoms in dark clouds.

us that dense dark clouds will have a fractional ionization of about 10^{-8}. This implies that a cloud whose collapse is retarded by the magnetic field and having a number density of $10^{11}\,m^{-3}$ will survive about 400 000 years. So, although magnetic fields do provide support against gravity for dense clumps, this support is relatively short lived, and the collapse cannot ultimately be prevented.

5.13   The inference of the ionization rate in dark clouds

Do the cosmic rays providing ionization in diffuse clouds actually penetrate the dark regions of thicker clouds? If not, then the assumption of $10^{-17}\,s^{-1}$ that we have used for that rate in our estimate for the survival time of a cloud with a number density of $10^{11}\,m^{-3}$ may be suspect. In fact, there are ways of deducing the ionization rate in dark clouds. These also depend on interstellar chemistry.

The most direct method involves the measurement of the atomic hydrogen abundance in the coldest parts of a dense molecular region. This abundance can be measured because the cold H atoms in a molecular region absorb 21 cm radiation emitted by hotter H atoms behind the cloud. Figure 5.14 shows the chemistry that controls the H production in a dark molecular region. For every formation of H_2^+ following a cosmic ray interaction with H_2 several H atoms are produced. The H atoms are removed in collisions with grains; this removal occurs at a reasonably well-known rate. This network tells us that the equilibrium abundance of H atoms is directly proportional to the cosmic ray induced ionization rate. From measurements of the H abundances in various *dark* clouds the ionization rate has been estimated to be about $10^{-18}\,s^{-1}$, or roughly a factor of 10 lower than the rate in most *diffuse* clouds. The dependence of the ionization rate on cloud thickness is also consistent with many of the ionizing cosmic rays having speeds around 4 per cent that of light if the cosmic rays enter the clouds unhindered by magnetic effects. The adoption of the lower ionization rate in dark clouds reduces the level of ionization by a factor of 3, and consequently also reduces by a factor of 3 the period of time that a magnetic field can retard the collapse of a cloud.

5.14   Another way to find the fractional ionization in dark clouds

The fractional ionization is clearly an important factor in determining the cloud's evolution. We have, so far, estimated it in a way that requires us to

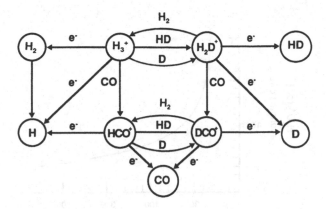

Fig. 5.15. The chemistry that establishes the DCO^+ to HCO^+ abundance ratio in dark clouds.

specify the cosmic ray induced ionization rate in a chemical model. There are, however, other methods we can use to arrive at the fractional ionization and in this section we consider one that does not require the ionization rate to be known. Upper bounds to the fractional ionization can be obtained by measuring the abundance ratio of DCO^+ and HCO^+ if the cosmic deuterium abundance is specified. Figure 5.15 shows schematically the chemistry that establishes the DCO^+/HCO^+ ratio.

Deuterium is heavier than hydrogen; hence, a molecular species containing deuterium vibrates less rapidly than the corresponding species that contains hydrogen. For purely quantum mechanical reasons even the ground level of a molecular species containing deuterium has a lower energy than the ground level of the corresponding species containing hydrogen. (Recall that even in its lowest energy vibrational level a molecule is vibrating and has a nonzero vibrational energy; that energy is lower for a deuterated species than its corresponding protonated species.) At low temperatures the abundances of low energy species increase rapidly with respect to higher energy species.

The cosmic abundance ratio of deuterium to hydrogen is thought to be around $1:30\,000$. (We shall describe one means of measuring that ratio in the next section.) In dark molecular clouds nearly all deuterium is in HD and nearly all hydrogen is in H_2 implying that the HD to H_2 abundance ratio is around 1–2 times the cosmic abundance ratio of deuterium to hydrogen.

We shall assume throughout this paragraph (though only this paragraph!) that the electron, CO, and atomic deuterium abundances in dark molecular cloud are very small. Then reactions of HD with H_3^+ and HCO^+, opposed by reactions of H_2 with H_2D^+ and DCO^+, control the H_2D^+/H_3^+ and DCO^+/HCO^+ abundance ratios. If the temperature were high then the energy differences between H_2D^+ and H_3^+ and between DCO^+ and HCO^+ would be relatively unimportant, and the H_2D^+/H_3^+ abundance ratio and the DCO^+/HCO^+ abundance ratio would both be equal to the HD/H_2 abundance ratio.

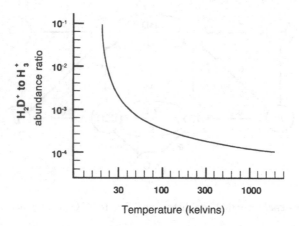

Fig. 5.16. The temperature dependence of the H_2D^+/H_3^+ abundance ratio. Low electron, CO, and atomic D abundances have been assumed. If they were taken to be higher, the curve would not climb so high at low temperatures.

However, at the low temperatures of dark clouds the difference in the energy of a deuterated ion and its protonated counterpart becomes important. The H_2D^+/H_3^+ and DCO^+/HCO^+ abundance ratios climb to somewhere between about 1:10 to 1:100. This behaviour is illustrated in Fig. 5.16.

The actual chemistry is more complicated than the simple one described in the previous paragraph, and the increases in the H_2D^+/H_3^+ and DCO^+/HCO^+ ratios as the temperature decreases are less pronounced than indicated in Fig. 5.16. The temperature dependences of those ratios in the presence of electrons, CO, and atomic deuterium are, however, well understood. On the assumption that the deuterium/hydrogen abundance ratio is the same in all sources, we can use the measured DCO^+/HCO^+ ratios for clouds whose temperatures are known (from the analysis of the strengths of a number of emission features) to determine upper limits for the abundances of CO and electrons. The upper bound for the electron abundance is consistent with the ionization rate being in a range around $10^{-17}\,s^{-1}$.

5.15   Diagnosis of conditions following the Big Bang using present day molecules

As our final discussion in this chapter we turn to two topics related to the Early Universe, which was treated in Chapter 4. We had to wait until we had considered interstellar chemistry in some depth to provide the background necessary to approach those subjects.

The cosmic deuterium/hydrogen ratio is one of a small number of data from which the long term future of the Universe can be inferred. As described at the beginning of Chapter 4, the nuclear burning in the Universe when its temperature was still as high as about 1000 million kelvins produced helium from hydrogen and also synthesized deuterium. The amount of deuterium produced in an expanding medium is sensitive to the density and the expansion timescale as the temperature of the medium passes through a temperature range around 10^9 kelvins. The density and rate of expansion at any era also determine whether the medium will collapse back upon itself eventually or expand without limit. Hence, the cosmic deuterium/hydrogen abundance ratio is a measure of the extent to which the Universe will expand in the future.

As will be described in Chapters 8 and 10 nuclear burning also occurs in stars, and so the ejection of stellar processed matter into the interstellar medium has altered the present day Galactic deuterium/hydrogen ratio from its cosmological value. It is generally believed that the mixing of stellar processed material into the interstellar medium decreases the deuterium/hydrogen ratio from the primordial value because deuterium is burned in stars. However, many different types of nuclear burning occur in the huge variety of different classes of stars, and the details of the mass return and burning during the mass return stages of stellar evolution are imperfectly understood. Hence, conceivably, the Galactic deuterium/hydrogen ratio may even be enhanced relative to the cosmological value.

It is possible to infer the primordial value of the deuterium/hydrogen abundance ratio and, hence, the information that it gives concerning the future of the Universe, only if the effect of stellar processing on the Galactic ratio can be determined. In fact, the Galactic value of the deuterium/hydrogen ratio almost certainly varies, as some parts of the Galaxy have been less processed by stars than others. For instance, the Galactic Centre region is a site of continuing star formation while the outer regions of the Galaxy are more quiescent. We need a means of determining the deuterium/hydrogen abundance ratio in a variety of places throughout the Galaxy so that we can determine the cosmological outlook.

The only known such means feasible with currently existing instrumentation involves the measurement of the abundance ratio of species like DCO^+ and HCO^+. The abundance ratios of deuterated and protonated forms of other species, especially DCN and HCN, are also used, and the chemical control of those ratios is similar to that of the DCO^+/HCO^+ ratio. At the end of the previous section we stated that the observed DCO^+/HCO^+ ratios vary with temperature in a way that is consistent with the electron and CO abundances being small and the deuterium/hydrogen abundance ratio being constant. This constant deuterium/hydrogen ratio can be derived from the observed DCO^+ to HCO^+ data. However, observations of DCO^+ and HCO^+ and other deuterated and protonated pairs in a greater variety of sources may reveal deviations of the temperature dependence from that expected for low electron and CO abundances and a constant deuterium/hydrogen ratio. These deviations may prove to be due to variations in the deuterium/hydrogen ratio. However, that conclusion could only be established after other species, such as OH, were used to measure the cosmic ray induced ionization rates and the electron number densities and after the CO abundances were determined from observations of many different CO emission features. With hard work, a very important insight into the fate of the Universe would be obtained.

Interstellar molecules have played an important historical role in obtaining other fundamental data about the evolution of the Universe. The first detected sign that the Universe is filled with 'black body' radiation remnant (see Chapter 4) from the Big Bang was obtained from the optical absorption features produced by CN molecules against background stars. The data were interpreted by G Herzberg in 1950 as implying the existence of a radiation background, but, unfortunately, its origin was not identified for one and a half decades. Once the identification was made, the data provided useful information about the background radiation at a frequency that was then inaccessible with other observational methods. The CN molecule shows several separate optical spectral features in absorption against a background star. One is due to CN in its lowest rotational level (i.e. the nonrotating level) and the others arise from CN in its first excited rotational level. The upper rotational level is excited by the absorption of photons from the 'black body' background, and the populations of the lowest two rotational levels give a measure of the strength of the background radiation. The CN measurement is consistent with other direct detections of the radiation, providing some comfort to students of cosmology, a field filled with great uncertainty.

Selected references

Fiedler, R A and Mouschovias, T Ch: 'Ambipolar diffusion and star formation: Formation and contraction of axisymmetric cloud cores II. Results', *Astrophysical Journal*, vol. 415, p.680 (1993).

Friberg, P and Hjalmarson, Å: 'Molecular clouds in the Milky Way', in *Molecular Astrophysics – A Volume Honouring Alexander Dalgarno*, ed. T W Hartquist, Cambridge University Press, Cambridge (1990).

Hartquist, T W and Morfill, G E: 'Cosmic ray diffusion at energies of 1 MeV to 10^5 GeV', *Astrophysics and Space Science*, vol. 216, p. 223 (1994).

Lepp, S and Dalgarno, A: 'The ionization rate in dense interstellar clouds', in *Astrochemistry – IAU Symposium 150*, eds. M S Vardya and S P Tarafdar, D Reidel Publishing Company, Dordrecht (1987).

Millar, T J: 'Chemical modelling of quiescent dense interstellar clouds', in *Molecular Astrophysics – A Volume Honouring Alexander Dalgarno*, ed. T W Hartquist, Cambridge University Press, Cambridge (1990).

Millar, T J and Williams, D A (eds.): *Dust and Chemistry in Astronomy*, Institute of Physics Publishing, Bristol (1993).

Roth, K C, Meyer, D M and Hawkins, I: 'Interstellar cyanogen and the temperature of the microwave background radiation', *Astrophysical Journal Letters*, vol. 413, p.L67 (1993).

Spitzer, L Jr: *Physical Processes in the Interstellar Medium*, John Wiley and Sons, New York (1978).

van Dishoeck, E F: 'Diffuse cloud chemistry', in *Molecular Astrophysics – A Volume Honouring Alexander Dalgarno*, ed. T W Hartquist, Cambridge University Press, Cambridge (1990).

Star formation

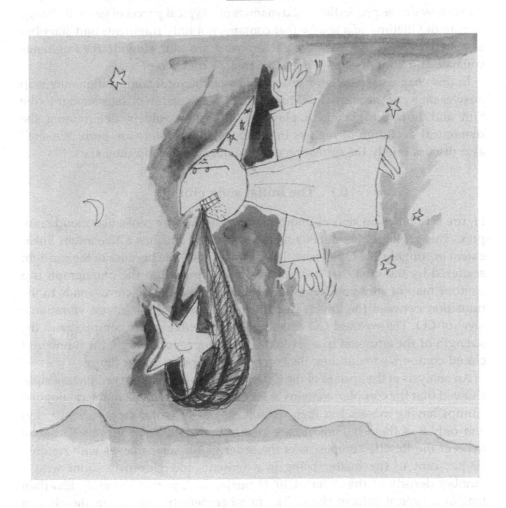

Stars are being born in the Galaxy at a rate of about one per year. Many of the newly born stars will survive for over 10 billion years, a time comparable to the present age of the Galaxy. Others will evolve and burn out, in some cases with violent explosions called supernovae (see Chapter 10), on time-scales of roughly 1 million years. As we saw in Chapter 2, this is a fairly short time in astronomical terms; it is 1 A-day. It is only about a hundredth of the Galaxy's rotation period.

Stars form in interstellar clouds, and the story of a star's birth is also the tale of the evolution and collapse of a part of a cloud. A cloud can be a fairly long lived stable object. Typical ages lie in the range of 10–100 million years. The central questions are these: what mechanism triggers the collapse of a part of a cloud? Once infall begins, on what timescale does significant evolution take place? Does fragmentation occur? Is some type of quasi-stable intermediate structure (consisting, perhaps, of a cluster of gaseous, dusty fragments and stars born in them) established? If so, how long does such a structure survive? How many generations of young stars are created in it before its dissolution? How do the properties and dynamics of a typical parcel of gas in it change during the lifetime of a source that contains not only fragments and stars but also stellar winds that blow the gas and dust around? How do the fragments collapse and why do stars possess the masses they have?

In this chapter, we tell how observations of chemical composition may help answer many of the above questions about the formation of stars similar to the Sun and how the chemistry plays an important role in determining the dynamical evolution of regions in which solar-type stars are born. We shall also discuss briefly the formation of more massive and brighter stars.

6.1   The initial state clumps

Figure 6.1 shows an optical photograph of the Rosette molecular cloud complex. This region is about 5000 light years away and has a maximum linear extent of roughly 300 light years. The complex is visible because of the starlight scattered by the dust that it contains. Superimposed on the photograph is a contour map of emission from CO. The emission detected corresponds to the transition between the lowest two rotational levels of the lowest vibrational level of CO. The weakest CO emission is at the edge of the complex, and the strength of the emission in a given direction is proportional to the number of closed contours surrounding the corresponding point on the figure.

An analysis of the spectra of the CO emission along all observed lines of sight showed that the complex contains at least 86 clumps, with 33 of the catalogued clumps having masses less than 100 solar masses (1 solar mass $= 2 \times 10^{30}$ kg) and only 8 of them having masses greater than 1000 solar masses. The total mass of the Rosette complex is of the order of 10^5 solar masses with roughly 10 per cent of the matter being in a tenuous interclump medium with a number density of the order of 10^7 H nuclei m^{-3}, perhaps slightly less than that of a typical diffuse cloud. The number densities of gas in the clumps vary somewhat, but 10^9 molecules m^{-3} is a typical value.

Observations like those made of the Rosette complex show that most clumps in large clouds and complexes have fairly similar number densities and are translucent at optical wavelengths. Typically, only about 10 per cent of the optical light incident on a clump penetrates to its centre, the remainder being

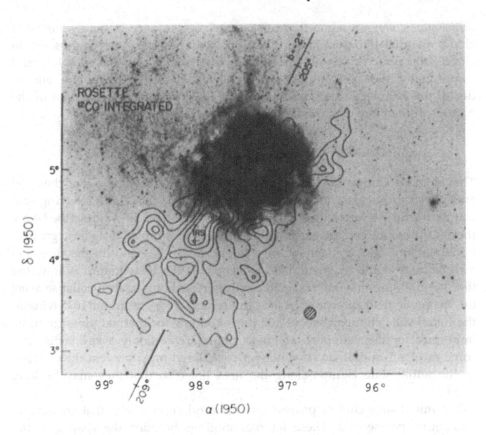

Fig. 6.1. The Rosette cloud complex. A contour map of CO emission overlays a visual photographic negative of the complex. The weakest detected CO emission originates at the edge of the complex where too little stellar light is reflected to be registered on the photographic image. The point marked IRS is the site of an infrared source due to the presence of dust heated by an embedded star. The circle with diagonally oriented slashes shows the angular resolution at which the CO observations were made. (From L Blitz in *Physical Processes in Interstellar Clouds*, eds. G E Morfill and M Scholer, D Reidel Publishing Company, Dordrecht (1987).)

scattered and absorbed by dust. The variation from clump to clump of the number of hydrogen nuclei in a column of fixed cross section going through the centre of the clump is rarely more than a factor of 2 or so. This is a small variation given that the masses of the clumps vary by a factor of about 100.

There must be clear chemical and physical reasons why the clumps with such a wide range of masses have such a narrow range in opacity. It also seems likely that the clumps with this range of optical depths are long lived; otherwise, they would not be so prevalent. Yet it is clear that stars form in complexes like the Rosette, implying that gas in these clumps almost certainly collapses eventually, since they are the only identifiable structures having densities between those of the more tenuous interclump regions and those of active star forming regions. We shall subsequently refer to clumps that allow of the order of 10 per cent of the visual light incident on them to penetrate to their centres as *initial state clumps*, a name that reflects their importance in the evolution of cloud material into star forming regions. Some initial state clumps are in clouds that are considerably less massive than the Rosette complex, and some small clouds contain only one initial state clump. However, the remainder of this chapter is applicable to any of those clumps, independent of the natures of the more extensive environments in which they exist.

6.2 Why are the initial state clumps stable?

That so many clumps possess roughly the same opacity implies that chemical and physical processes must act together to ensure that clumps having that common property are long lived. In Sections 4.2 and 5.10 we described how thermal pressure and magnetic fields in gas can support it against gravitational collapse. But a magnetic field supports the gas only in the direction *perpendicular* to the magnetic field direction. If the temperature is low, the thermal pressure may be too small to support a clump against collapse *along* the magnetic field direction. In fact, the typical temperature of 10 kelvins in the initial state clumps is about 100 times too low for thermal pressure to be important for their support. Magnetic fields are certainly a key factor, but some mechanism that we have not yet considered must prevent the collapse of the initial state clumps along the field lines, so that they may be long lived.

The initial state clumps possess internal random velocities that are several kilometres per second. These internal motions broaden the shapes of the spectral lines in which the observed molecular radiation is emitted. The motions are thought to consist of waves in the gas, and are probably important for the retardation of collapse along the magnetic field lines. As anyone who has tried to stand on a shallow ocean bottom knows, the reflection of water waves transmits force to the body. Reflections of gaseous waves occur continuously inside an initial state clump containing many strong waves, and these reflections transmit a force that pushes clump material away from the clump centre towards which gravity pulls.

How do these waves arise? Their origin is not clearly understood. Perhaps waves arise in the weaker of the clump–clump collisions that do not alter the clumps' structures too violently, or in the action of young stars on the material around them. Without a source of energy, waves decay. On Earth, the wind continually powers ocean waves, but in the absence of the wind, we would see an ocean wave decay in height as it propagates. Whatever may be the origin of the waves in interstellar clumps, it must be acting frequently enough to ensure that the waves do not decay. In initial state clumps, the magnetic field affects the waves. For very low levels of ionization the waves consist purely of disturbances in the magnetic field and in the ions and electrons that move in the magnetic disturbances. The neutral gas is unaffected, so the ions and neutrals have quite different velocities. This means that friction damps the waves strongly. For very high ionizations, however, the frictional coupling between the ions and the neutral gas particles is so great that the neutral gas is also disturbed by the waves and the ion and neutral components of the gas have velocities that are almost the same. In the more highly ionized material, therefore, the waves affect more mass and they decay far more slowly than do waves in lowly ionized gas. Figures 6.2(a), (b), and (c) illustrate these effects schematically.

Hence, if waves in an initial state clump help to prevent it from collapsing, the length of time that the clump will survive depends on the fractional ionization in it. In diffuse clouds, through which most optical radiation passes without being absorbed, almost all gas phase carbon is ionized, while most other species are neutral. The fractional ionization is, therefore, about 10^{-4} (see Section 5.6). However, at dark cloud centres where optical radiation does not penetrate fractional ionizations are about 10^{-7}, or less, and are maintained by cosmic rays (see Sections 5.5 and 5.11). The initial state clumps are translucent, and their fractional ionizations lie between these two extremes. The initial state clumps are just opaque enough that much of the carbon in them is in CO, so that we can detect them easily in maps of CO emission. They must also be sufficiently transparent that their fractional ionization is high and, consequently, the waves which help prevent the collapse of the clump decay very slowly.

Indeed, the initial state clumps are just opaque enough for carbon to be mostly in CO at the centre while nearly all gas phase sulphur and silicon atoms are ionized. Carbon is mostly in CO in part because it, like H_2, can shield itself against the destructive radiation. Most of the rest of the oxygen not taken up in CO is in other neutral species (see Section 5.5). In the initial state clumps the intensity of radiation is great enough to destroy sulphur- and silicon-bearing molecules faster than they form and to ionize atomic sulphur and silicon to form S^+ and Si^+. However, an increase by a factor of 2 or 3 in the optical depths at visual wavelengths would lead to the conversion of most of the sulphur and silicon to neutral molecules, with a consequent large drop in fractional ionization. The behaviour of the ionization structure as a function of cloud depth is depicted in Fig. 6.3. The decrease in fractional ionization is accompanied by an increase by many orders of magnitude in the rate at which waves damp, and an associated onset of gravitationally induced collapse along the field lines. Hence, clumps which have optical depths somewhat greater than those of the initial state clumps are passing through a transitory state until some other means of arresting their collapse takes effect.

6.3   The collapse from the initial state; core cluster formation

What mechanism triggers the collapse from the initial state? The answer is not known. However, a collision between two clumps will lead to compression and an increase in the optical depth of each and an associated fall in fractional ionization, more rapid wave decay, and collapse along the field lines. Hence, the strongest clump–clump collisions may drive the onset of collapse.

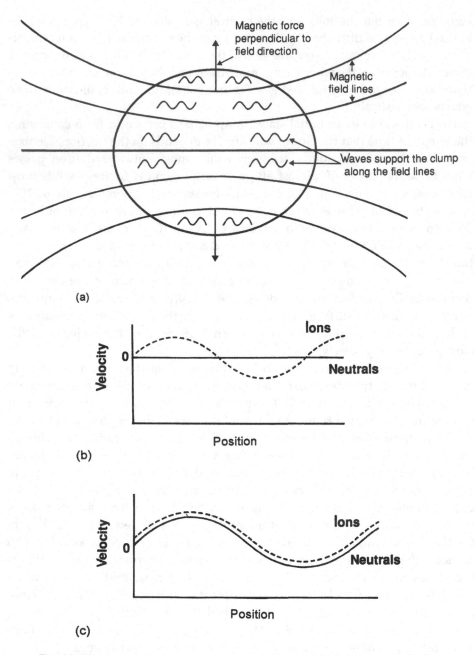

Fig. 6.2. Wave support of clumps and waves in low and high fractional ionization cases. (a) A clump supported by the magnetic field and waves. (b) Wave in the low fractional ionization case. The wave causes the ions to move but not the neutrals. The relative velocity between ions and neutrals is large and the wave decay is rapid. (c) Wave in the high fractional ionization case. The frictional coupling between ions and neutrals is high, and the wave causes the ions and neutrals to move with a small relative velocity. Wave decay is slow.

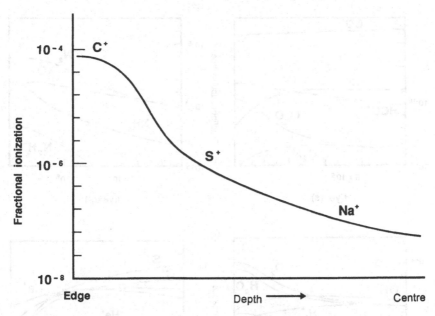

Fig. 6.3. Schematic diagram of the fractional ionization as a function of depth into a clump. The ions C^+, S^+, and Na^+ are indicated at the points where they are the dominant ions. The rapid drop in fractional ionization with the increase in the number in a column results in rapid decay of waves and collapse along field lines.

Once infall begins, it probably proceeds in an unhindered fashion on a time-scale equal to what is called the *free-fall timescale* which is calculated with the use of the theory of gravity. The free-fall timescale, t_{ff}, is roughly one million years for a cloud with an H_2 number density of $10^9 \, m^{-3}$; t_{ff} decreases with increasing density as the inverse of the square root of the number density. A free-fall timescale of one million years should be compared with two important chemical timescales. One is the timescale required to ionize an amount of H_2 comparable to the gas phase abundance of oxygen and carbon (see Section 5.5). This is the shortest timescale on which the nature of the chemistry can be entirely changed through ion–molecule reactions, and is about several hundred thousand to one million years. Some parts of the chemistry (e.g. the fractional ionization) adjust much more rapidly than others (e.g. the gas phase production of NO and CN), but about one million years is a reasonable timescale to associate with the chemistry. Hence, dynamical and structural changes can occur on timescales short enough that some signatures of past physical conditions remain in the chemistry. In principle, these chemical signatures of the past can be exploited in determining the dynamical history of a region.

The other important timescale is that associated with freeze-out of gas phase species onto grains (see Section 5.7). This can be estimated from known grain properties (derived from studies of the absorption of starlight by clouds) and the speeds of molecules in the gas. The freeze-out timescale is roughly $3 \times 10^{15}/n_H$ years where n_H is the number density of hydrogen nuclei in the

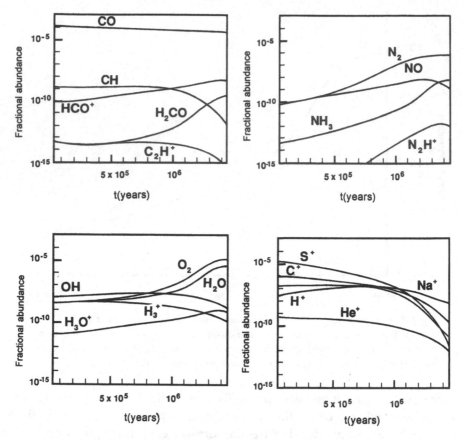

Fig. 6.4. Fractional abundances during the collapse of a clump. (From T W Hartquist, J M C Rawlings, D A Williams and A Dalgarno, *Quarterly Journal of the Royal Astronomical Society*, vol. 34, p.213 (1993).)

gas (m^{-3}). Note that, for a number density n_H of $10^{10}\,m^{-3}$ (or greater) the freeze-out timescale of about 3×10^5 years (or shorter) is less than the free-fall timescale.

In Fig. 6.4 the calculated fractional abundances of some species at the clump centre are given as functions of the time after the onset of the collapse. The figure gives rather a lot of information and is included to give some impression of the degree to which chemical models of astronomical objects must be developed in order for astronomers to exploit fully all of the observational data that potentially can be obtained in order to study the properties of various astronomical sources. In this calculation the initial state clump had a starting number density of 10^9 hydrogen nuclei per m^3 and the extent of the clump is such that about 2/5 of the optical radiation falling on the clump reaches the centre. After the infall had continued for a million years the number density and fraction of incident optical light reaching the clump centre were $2.65 \times 10^9\,m^{-3}$ and 0.15 respectively. After 1.44×10^6 years the number density was held constant at $2.69 \times 10^{10}\,m^{-3}$. The zero time fractional abundances were those that are computed for the initial state clumps. Notice that at early times S^+ was abundant, but that as the clump became denser and its centre darker Na^+ became the most abundant ion. The S^+ abundance became less than the Na^+ abundance after the collapse had proceeded for roughly one million years. Because it is self-shielding against destroying radiation CO was abundant at all times. However, as the centre became darker, there were increases in the abundances of species that were destroyed in the initial state clumps by radiation. H_2O is a good example of such a species. A slight drop in the CO abundance is noticeable at late times; it is due to accretion and freeze-out onto grains, a point of substantial importance (see Section 6.5).

Unimpeded collapse, which may lead to fragmentation, results in some star formation. However, the first and subsequent generations of stars formed in a clump act in a manner that arrests the free-fall of much of the clump material. Young stars have powerful winds. When it is young, a star similar to the Sun loses as much as a millionth of its mass in a single year through a stellar wind with a speed of several hundred kilometres per second. The winds of stars in a fragmented collapsed clump act back on the clump material and support some of it against the gravity.

Following free-fall collapse, an initial state clump probably evolves to produce an object similar to the star forming region called Barnard 5 (or B5), a map of which is shown in Fig. 6.5. It contains five identified fragments, called cores, surrounded by an intercore medium. It also possesses four stellar objects which heat the nearby dust; the hot dust emits infrared radiation. Each stellar object has a mass comparable to that of the Sun and produces an energetic wind. The average hydrogen number density and mass of each core are of the order of $10^{10}\,m^{-3}$ and 10 solar masses, respectively. The intercore medium is about a tenth as dense as a core, and it and all the cores have temperatures of 10–30 kelvins. We will refer to an object like B5 as a 'core cluster'. The stars in B5 will become solar-like after they have evolved further. We shall consider the formation of the more luminous stars having masses above about 4–8 solar masses only in the final section of this chapter. The trigger of solar-like star formation is probably gentler than that of the massive stars.

The stars, except for IRS1, in B5 are not in cores. This might seem strange, since the stars are born in the cores. However, once stars condense they are no longer coupled by friction and by the magnetic field to the motion of other matter, as are the gas and dust outside the stars. Thus, the stars move freely, unlike the core material, in the gravitational field and become displaced from the cores in which they formed.

Figure 6.6 shows a stellar wind sweeping around the boundary of a core and ablating gas and dust from it. The nature of the wind–core interaction at the interface between the now fast flowing wind gas, which consists mostly of H^+ and He^+, and the dense molecular core gas is highly uncertain. It is likely that the flow along the interface becomes turbulent just as the flow along a bumpy aeroplane wing would be. If so, the turbulent motions extend into the core to a depth of the order of a few to 10 per cent of the core size. These turbulent motions raise the core material out to the interface as it

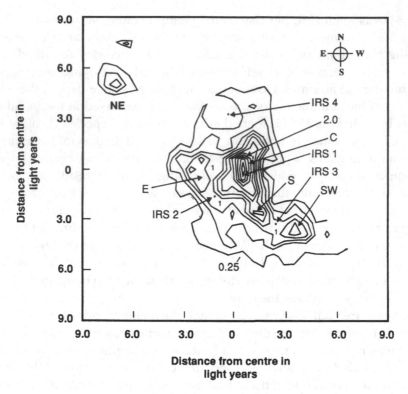

Fig. 6.5. Contour map of CO emission from Barnard 5. B5 is a star forming region containing five identified fragments, called cores, labelled NE, E, C, S, and SW. The strength of the CO emission along a line of sight is proportional to the number of closed contours surrounding the corresponding point on the map. The cores are associated with the regions surrounded by the most contours and have higher densities than the intercore regions. Four young stellar objects appear as infrared sources labelled IRS1, IRS2, IRS3, and IRS4. The object is about 1000 light years distant and is several light years across. (From P F Goldsmith, W D Langer and R W Wilson, *Astrophysical Journal Letters*, vol. 303, p.L11 (1986).)

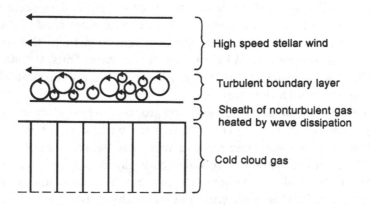

Fig. 6.6. A stellar wind–core boundary layer.

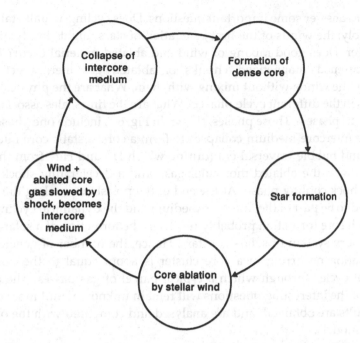

Fig. 6.7. A schematic representation of the cyclic physical history of a parcel of gas in a core cluster.

becomes progressively accelerated, probably in a fairly smooth manner, in the direction parallel to the interface. Ablation destroys the cores and is, therefore, one means of inhibiting further star formation, since without a core, no star can form. As the wind sweeps past a core it becomes loaded with molecular gas and perhaps becomes well mixed with it. All of this activity generates disturbances that heat an outer sheath of the molecular core.

6.4   Chemical–dynamical cycling in regions of solar-like star formation

The mixture of stellar wind and the molecular gas that it has eroded from the core blows a bubble much like that blown by the wind of a young galaxy (see Section 4.3). The mixture is slowed and heated by passing through an inward facing shock near the bubble edge. The gas cools, and reaches a physical state similar to that of the ambient intercore medium. In fact, the intercore medium which condenses to form cores is continually replenished by the deceleration and cooling of the wind-ablated gas mixture. Thus, a parcel of gas continuously cycles through the intercore medium, the cores which form from it, and the moving wind-ablated gas which is decelerated to restore the intercore medium. A dynamical cycle that might typify those through which gas in a core cluster passes is illustrated in Fig. 6.7.

The timescales associated with the different phases of the cycle are highly uncertain. One of the goals of studying the chemical composition of objects like B5 is to answer some important questions. Does cycling actually take place? Conceivably, the winds of the first generation of stars might simply disperse a core cluster. Does good mixing of wind and ablated material occur? Perhaps, highly flattened magnetized chunks of ablated gas merely get pushed around by the winds without mixing with them. What are the physical properties of gas in the different cycle phases? What are the timescales associated with the different phases? Those phases, shown in Fig. 6.7, include one phase during which the intercore medium collapses to form a core, a static core interval, an ablation and bubble traversal era (during which H^+ and He^+ from the stellar wind mix with the ablated molecular gas), and a decelerating shock heating and postshock cooling phase. At the end of the postshock cooling the mixture is assumed to be part of the intercore medium and the dynamical cycling begins again. Each core formation probably results in the formation of a solar-like star or of a binary system of solar-like stars. Hence, the number of generations of star formation occurring in a core cluster is about equal to the number of dynamical cycles through which a typical parcel of gas passes. The answers to many of the interesting questions will remain unknown until more observational results are obtained, and are analysed and compared with the output of chemical models.

The chemistry of the gas varies from phase to phase of the cycle. During the collapse of intercore gas to form a core and during the lifetime of the core, gas phase chemistry proceeds. At the same time atoms, ions, and molecules containing elements heavier than helium stick to any grains with which they collide. The way in which molecules are processed on grains is not clearly understood but CO probably remains as CO, oxygen-bearing species such as O and OH probably react with hydrogen on the grain surfaces to form water ice, and some nitrogen-bearing species most likely react with hydrogen to form ammonia ice. How these ices are removed from the grain surfaces is unclear. Infrared absorption observations towards stars embedded in cores confirm that ices exist on the grains, but only when the cores are quite optically thick. Radiation may help to return these ices to the gas in those parts of a cloud where the radiation can penetrate fairly freely. Hence, during the core collapse and static core phases, ice formation on grain surfaces occurs in parallel with the gas phase chemistry.

During the ablation and mixing phase, the molecules from the core gas are mixed with H^+ and He^+ from the stellar wind. Although H^+ and He^+ react with H_2 only slowly at low temperatures, the chemistry of other molecules is affected substantially by these ions. When the gas passes through the decelerating shock near the bubble edge then the ices are released from the grain surfaces as the collisions of hot gas particles with the grains sputter the ices. Thus, H_2O, CO, NH_3, and other molecules on the grain surfaces are injected into the shocked gas.

We described in Section 5.8 how shock heating of a gas can result in a rich chemistry which removes many of the atomic species in the gas phase and creates molecules containing as much hydrogen as possible. For instance, this gas phase shock chemistry is a source, in addition to the sputtering from grains, of H_2O formed by the reactions of O and OH with hydrogen. Due to the grain sputtering and to the gas phase shock chemistry, the postshock wind-ablated material mixture contains a variety of molecules, such as H_2O and NH_3, having much higher abundances than expected in static cold clouds. However, the optical depth of the core cluster in the vicinity of a shock and the general intercore medium is probably low enough that many of these molecules are destroyed by photoabsorption due to starlight. CO, however, as discussed at the end of Section 6.2, is shielded and is not destroyed by starlight in these environments.

However, if the mixture contains sufficient H^+ and He^+, some of the CO molecules will be broken down in the cool postshock gas. He^+ ions react with CO molecules, producing C^+ ions, which recombine with electrons to form neutral carbon atoms. This may be an important source of neutral atomic carbon in the intercore gas if mixing occurs. The neutral atomic carbon, depending on the core collapse timescale and the cosmic ray ionization rate of H_2 (which determines the timescale required to convert C to CO), may survive through the core collapse phase and in the core. Neutral atomic carbon emission has been detected towards B5 at a level that requires a much higher atomic carbon abundance (about 10 per cent of the carbon must be in the form of neutral atomic carbon) than is compatible with standard dark cloud low-temperature gas phase chemistry without He^+ injection. *The neutral atomic carbon emission may imply that mixing of wind and ablated material and repetition of the dynamical cycle are important parts of the dynamics by which the core clusters evolve to form low-mass stars.* Detailed mapping of CO, C^+, and C emissions in the vicinities of the core–wind interfaces in core clusters would give considerable insight into the nature of the interaction of diffuse fast flows and gaseous obstacles. Such interactions are important for the appearances and evolutions of a huge variety of astronomical sources but are very poorly understood; diagnostic studies of the core cluster interfaces would, therefore, contribute significantly to the study of a basic astronomical process of general and profound relevance.

6.5   The infall of a core to form a star

Radio emission from NH_3 molecules is often used to study the detailed structures of the cores. Figure 6.8 is an NH_3 emission map of a core containing a recently born star. The fractional abundance of NH_3 in cores is typically in the range of 10^{-8}.

An interesting feature of the NH_3 emission line profiles is their narrow width. The dashed curves in Fig. 6.9 show a typical NH_3 emission line shape. The height of the curve is proportional to the fraction of the emission coming from molecules moving at the corresponding velocity. There is a velocity–frequency or velocity–wavelength relationship: gas moving towards the observer emits radiation that appears at a higher frequency than that

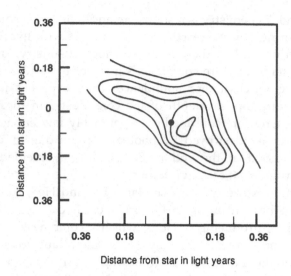

Fig. 6.8. NH$_3$ contour map of a dense core. The strength of the NH$_3$ emission is proportional to the number of closed contours surrounding a point. The solid circle shows the position of a newly formed star detected by the infrared emission from the dust surrounding it. (From K M Menten and C M Walmsley, *Astronomy and Astrophysics*, vol. 146, p.369 (1985).)

from gas moving away from the observer, whereas radiation from gas that moves away appears to be at lower frequency and longer wavelength. This is analogous to the sound of an ambulance siren seeming higher in pitch as it approaches, and lower when moving away. Molecules in gas at a given temperature have a distribution of velocities which is wider when the temperature is higher. Hence, a high-temperature gas gives rise to a broad emission feature. Systemic motions such as collapse (in which gas nearer the observer moves away while gas further away, on the other side of the object, moves towards the observer) also affect the shape of the emission profile generally resulting in the profile wings being stronger. The observed NH$_3$ emission profiles for cores are very nearly the same as those expected from gas at the core temperatures (measured from comparisons of the ratio of the strengths of different emission features that depend on the gas temperature) *without* substantial systemic motions. *That is, the observed NH$_3$ emission profiles give no evidence for collapse;* the observed profile wings (dashed lines) are not enhanced as are those of the solid curves in the CH and H$_2$S panels in Fig. 6.9.

Why do the NH$_3$ data not indicate collapse? Perhaps many cores are stable, long lived products of the collapse and fragmentation process. Their estimated masses, temperatures, and magnetic field strengths are compatible with them being supported by the fields and the thermal pressure. In Section 5.10 we described how magnetic field pressure drives the motion of ions relative to the neutral gas and an associated weakening of the magnetic field, allowing collapse to occur. The magnetic field weakening and the collapse are slowed when the fractional ionization increases, as it may when the first generation

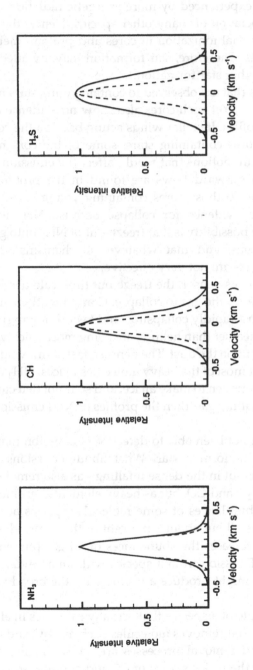

Fig. 6.9. Theoretically calculated emission profiles in a collapsing core. In each of the three panels the dashed curve represents the observed NH_3 emission profile. The rate at which species containing elements heavier than helium freeze out onto grains was adjusted so that the NH_3 would be removed from the gas phase before the gas started to enter the zone where infall is rapid; hence, the model NH_3 profile, shown by the solid curve in the first panel, is almost the same as the observed profile. The abundances of many species were followed as freeze-out occurred. The solid curves in the second and third panels show the calculated CH and H_2S emission profiles; the broad wings of these profiles are due to infall. (From T W Hartquist, J M C Rawlings, D A Williams and A Dalgarno, *Quarterly Journal of the Royal Astronomical Society*, vol. 34, p.213 (1993).)

of stars is born in a core cluster. The gas phase metals such as sodium are ionized by lower energy radiation than carbon. The radiation of young stars induces sodium ionization, and propagates through a core without absorption as substantial as experienced by more energetic radiation which induces ionization and dissociation of many other species. Hence, the birth of stars may increase the fractional ionization in cores and prolong their support, in part, by magnetic fields. Therefore, star formation may, by affecting the ionization structure, halt further star formation.

However, cores that are observed to contain young stars must be unstable to produce those stars. Yet such cores also show no evidence of collapse in their NH_3 emission profiles. If stellar winds acting back on the cores were arresting the collapse of cores containing stars, some of the core material should be driven outwards in motions that would affect the emission profiles; however, no signs of such outward flows are found in the profiles. Surely, collapse must be occurring in those cores containing young stars. Perhaps, the NH_3 profiles show no evidence for collapse because NH_3 is absent from the infalling gas. One possibility is that freeze-out of NH_3 onto grains has occurred in the infalling gas, and that whatever mechanisms act to return these molecules to the gas are not very effective.

As the gas becomes denser, the freeze-out timescale decreases more rapidly than the timescale for the gas to collapse. Consequently, while the densest gas in a core is the most rapidly collapsing (and fastest moving) it is also the gas in which elements, heavier than helium, are being most quickly removed from the gas and freezing out on the dust. The densest, fastest-moving gas in the collapse has, therefore, lost most of the heavy molecules in ices. NH_3 does not emit radio waves when it has frozen to make an ice, and so is not detected. If the NH_3 were all in ice in the infalling gas, then the profiles of NH_3 emissions would look just as they do.

So far, we have not been able to detect NH_3 emission that is clearly coming from gas falling to form a star. What about emissions from other likely molecules? Freeze-out in the dense infalling gas also removes other molecular species such as H_2O and CO, but as heavy elements become depleted onto the dust grains the abundances of some molecular species actually *rise* until the depletion removes all but about 1 per cent of the heavy elements; subsequent freeze-out then decreases the abundances of all gas phase species containing heavy elements. Emission from a species with an abundance that *increases* as freeze-out begins might produce a profile with the broad wings indicative of collapse.

CH is an example of a species that actually increases in abundance (at least, initially) as freeze-out removes molecules such as H_2O and CO. The dominant CH formation and removal processes are shown in Fig. 6.10. Reactions of CO with He^+ are the major source of C^+ and simultaneously the major cause of He^+ removal, implying that the C^+ formation rate depends on the He^+ formation rate but is insensitive to the CO abundance. CO is simply reformed if C^+ reacts with OH or H_2O. In depleted gas, the OH and H_2O abundances

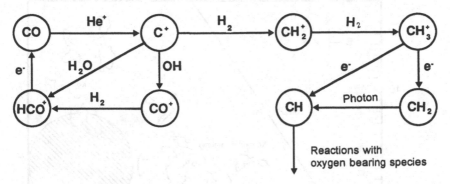

Fig. 6.10. CH formation and removal in dark clouds. The photodissociation of CH_2 is by radiation emitted during the slowing down of electrons that are ejected in the ionization of H_2 by cosmic rays.

drop, and the probability that C^+ will react with H_2 increases, resulting in an increased formation rate of CH; this by itself guarantees that the CH abundance rises as depletion by freeze-out occurs. However, the fact that CH is removed primarily in reactions with oxygen-bearing species with abundances that decrease with increasing depletion also acts to increase the CH abundance with increasing depletion by freeze-out.

Since the CH fractional concentration increases with increasing depletion, it should be more abundant in the denser infalling gas than in the static gas in a core. The CH line profile, obtained in a theoretical calculation, produced in a core in which the densest parts are infalling is shown by the solid curve in the middle panel of Fig. 6.9. In all three panels, dashed curves show a measured NH_3 emission profile. The theoretical CH, NH_3 and H_2S emission profiles are shown by solid curves in the figure. The H_2S abundance also increases with freeze-out in the infalling gas, though for somewhat different reasons than the CH. Perhaps, future observations of the appropriate emission features, selected following a consideration of the chemistry in depleted gas, will unveil the dynamics of star formation. CS emissions seem to indicate core collapse in some cases, but further study is necessary.

6.6   Regions of massive star formation

Stars with masses up to 50 or 60 times that of the Sun are also born in interstellar clouds. Chemical processes are undoubtedly important in controlling their formation, but we shall concentrate on the response of the molecules in the clouds to massive star formation. Stars with masses more than about ten times that of the Sun have sufficiently violent winds and experience strong enough explosive mass loss (as supernovae) to affect cloud structure on scales that are large compared to that of an individual core. The power of the mass loss from massive stars (with masses greater than 10 solar masses) almost certainly prevents regions in which massive stars form from surviving more than a single generation of stellar birth. The power of the wind of one of

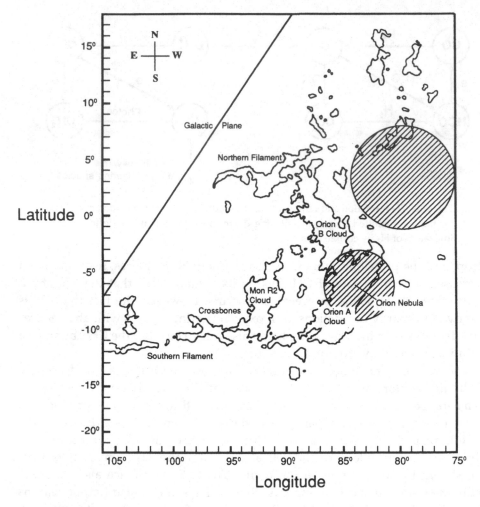

Fig. 6.11. A schematic diagram of the Orion molecular cloud complex. A number of dense molecular clouds, clumps, and filaments are distributed along the sky. Their positions are indicated in a variation of a standard system for measuring latitude and longitude on the sky. Three separate regions of high mass star formation lie in the Galactic disk, the midplane of which is indicated by a straight line in the upper left hand corner. The Orion Nebula, the region shown in Fig. 1.1, occupies a comparatively very small volume in the feature that is called the Orion A Cloud. (After R Genzel and J Stutzki, *Annual Review of Astronomy and Astrophysics*, vol. 27, p.41 (1989).)

the largest stars is roughly 100 times that of a solar-like star in even its windiest phase of evolution. The winds of the most massive stars have speeds that are up to ten times faster than those of more modest stars. The wind speed considerably affects the power exerted by a star on its surroundings.

The Orion cloud at a distance of about 1400 light years contains several regions of recent or ongoing massive star formation. Figure 1.1 is of the Orion Nebula, just one of those regions with massive stars. Those regions dominate the cloud's optical appearance. Figure 6.11 shows schematically the

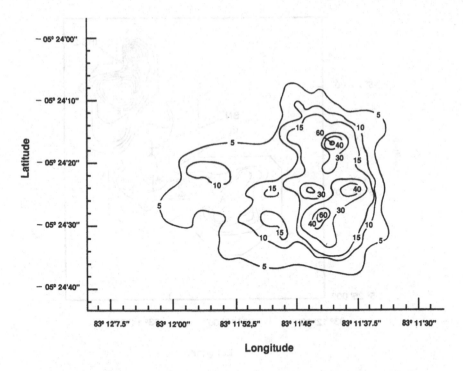

Fig. 6.12. The Kleinmann–Low Nebula mapped in infrared emission at 20 micrometres. At least one young massive star is embedded in the dusty molecular gas. It heats the dust which radiates the observed 20 micrometres infrared emission. Each vicinity surrounded by a great number of contours contains a dusty clump. The Kleinmann–Low Nebula lies behind the Orion Nebula as can be seen by comparing the coordinates of Figs. 6.11 and 6.12. (After D Downes, R Genzel, E E Becklin and C G Wynn-Williams, *Astrophysical Journal*, vol. 224, p.869 (1981).)

different regions in Orion. Associations of massive stars are identified by the cross-hatched areas on the map. Determinations of the stellar ages from the dynamical structure of each association and detailed comparisons of theoretical stellar evolution models with the observed stellar spectra imply that the associations are progressively younger as one moves along the Galactic plane to increasing longitude. The formation of the most massive stars appears to be sequential; further massive star formation is probably induced by the influence of the stellar winds on the surrounding cloud material. At the distance of the Orion cloud 10° of arc corresponds to about 250 light years; this is of the order of the separations between bright star associations in Orion, and the mass loss of the large stars must affect the cloud over distances that are many tens of times the dimension of a core cluster like B5.

The region of massive star formation in the Orion A Cloud has been studied extensively at many wavelengths. Figure 6.12 is a contour map of the infrared emission at a wavelength of 20 micrometres from that region which is called the Kleinmann–Low Nebula. The 20 micrometres radiation is emitted by dust

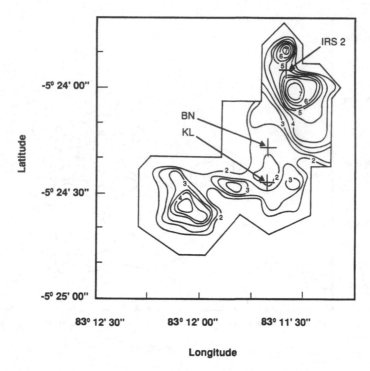

Fig. 6.13. H_2 emission in the Kleinmann–Low Region. A map of emission corresponding to transitions between rotational sublevels in the first excited and ground vibration levels shows that the emission is strongest near one of the clumps detected
by the 20 micrometre infrared emission from the dust. This clump is denoted IRS2
and is closely associated with a young massive star. The H_2 infrared emission near
that star is double peaked. (From S Beckwith, S E Persson, G Neugebauer and E F
Becklin, *Astrophysical Journal*, vol. 223, p.464 (1978).)

heated by the absorption of stellar radiation. Peaks on the contour map
correspond to sites of dusty clumps.

Figure 6.13 is a contour map of the infrared radiation emitted by H_2 in transitions from the first excited vibrational level to the ground vibrational level.
Note that the H_2 infrared emission possesses a strong double peak in the
vicinity of the source IRS2, a dusty clump either surrounding or very near
the brightest young star in the Kleinmann–Low Nebula. The two components
of the double peak are separated by about a tenth of a light year. The double
peak H_2 emission arises in a shock or shocks associated with the bubble
blown by the wind of the star near IRS2 (see the discussion of wind blown
bubbles in Section 4.3). Collisions between H_2 molecules can transfer energy
from the thermal motion of the H_2 to populate the first vibrational level of
H_2 when the temperature is in excess of about 1000 kelvins (see Section 3.10).
The properties of the H_2 emission from a number of levels imply that the
temperature in the gas in which it arises is around 2000 kelvins.

Probably the most intriguing problem concerns the very broad wings of the
H_2 emission profiles. They imply that the H_2 moves at velocities with a range of

over $100\,\mathrm{km\,s^{-1}}$. In a dense enough nonmagnetic region, a shock moving through the medium at a speed of $26\,\mathrm{km\,s^{-1}}$ or greater would destroy all H_2. Hence, the survival of H_2 in the presence of such violent motions is a puzzle. An attempt to solve this puzzle has been made by allowing for the effects of a magnetic field on the shock. This suppresses the high temperatures which cause H_2 destruction. However, the suggestions that have been made are not entirely convincing. Much remains to be learned about the dynamics of star forming regions, and the H_2 puzzle may be relevant to some very fundamental astronomical processes that require elucidation before we can understand a huge variety of astronomical sources in and outside of star forming regions.

Selected references

Bertoldi, F and McKee, C F: 'Pressure-confined clumps in magnetized molecular clouds', *Astrophysical Journal*, vol. 395, p.140 (1992).

Genzel, R and Stutzki, J: 'The Orion Molecular Cloud and Star-Forming Region', *Annual Reviews of Astronomy and Astrophysics*, vol. 27, p.41 (1989).

Hartquist, T W, Rawlings, J M C, Williams, D A and Dalgarno, A: 'The regulatory and diagnostic roles of chemistry in low-mass star forming regions', *Quarterly Journal of the Royal Astronomical Society*, vol. 34, p.213 (1993).

Nejad, L A M, Hartquist, T W and Williams, D A: 'Models of dense cores in translucent regions of low-mass star formation', *Astrophysics and Space Science*, vol. 220, p.261 (1994).

7

The Solar System at birth

The birth of the Sun, a low-mass star, was accompanied by the formation of planets, moons, asteroids, comets and meteors which – together with the Sun around which they orbit – form the Solar System. In the previous chapter we described in some detail the history of low-mass star formation from very low-density gas up to a number density of about $10^{13}\,\mathrm{m}^{-3}$. We then simply assumed that star formation occurs, and described the effects that the winds of young stars have on their environment. However, a great deal happens as the number density of a collapsing core increases from about $10^{13}\,\mathrm{m}^{-3}$ to about $10^{20}\,\mathrm{m}^{-3}$, which is roughly the value of the number density in the proto-Solar Nebula in the region where Jupiter was formed.

For instance, as we shall describe more fully below, up to densities of about $10^{13}\,\mathrm{m}^{-3}$ the magnetic field acts to prevent the spin-up of collapsing protostars. Above this density, however, spin-up increases as the collapse proceeds. The decrease of the efficiency of magnetic retardation of spin is due to the drop in the chemically controlled fractional ionization. The spin-up restricts the collapse rate of the gas in some directions but does not interfere with collapse parallel to the axis about which the spinning occurs; so a disk-like structure forms.

Much of what is known about the history of the proto-Solar System disk during the era of planet formation derives from the analysis of the chemical composition of meteoritic material. For instance, some meteoritic material may have been heated by lightning; if so, what does that imply about the chemical and ionization conditions in the primitive Solar Nebula? Another source of information is the present day composition of the Earth; what does its possession of water tell us about the chemical and physical conditions in the proto-Solar Nebula at the time of the planet's birth? What does our knowledge of the chemical composition of the comets tell us about the material from which they formed?

The atmosphere of each planet has a long history of chemical evolution, and the atmospheric chemistry controls the environments of the planets' surfaces. However, we will not describe the chemistries of planetary atmospheres, which would form interesting subjects for a separate book. As we have stated in the Preface two reasons for the omission of planetary atmospheres from consideration are: that the chemistries in them, unlike the chemistries of most of the astronomical environments that we treat in this volume, are affected to an important degree by reactions involving the simultaneous inter-action of more than two atoms, ions, or molecules; and that many books introducing astronomy to nonscientists contain material on the evolution of the chemical compositions of planetary atmospheres. We do touch briefly on the chemical composition of some planets just as they were forming but will not deal in depth with the problem of planet formation since some current theoretical models of planet formation indicate that the accumulation of solid bodies to make the dense metallic cores of planets should have required more time than is believed to have been available before the proto-Solar Nebula disappeared, partly by infall onto the Sun (i.e. some theoreticians would predict that planets don't exist!). While much of this chapter concerns the Solar System when it was still in the process of forming, we do include a discussion of an object which has remained a molecular source until the present: the Sun, which contains H_2 in its sunspots.

The topics discussed in this chapter should be viewed as a sample of subjects in the chemistry of the young Solar System.

7.1 The formation of the disk

The orbits of the Solar System's planets lie in a plane. If the proto-Solar Nebula had collapsed in a spherically symmetric fashion the orbits of fragments formed in it would be in many different planes. Hence, we can conclude that the dusty gaseous nebula from which the Solar System was born evolved through a disk-like configuration. Radio, infrared, and sometimes optical observations of nearby very young low-mass stars show that they often possess orbiting disks of dust and gas. Figure 7.1 shows an image of a disk associated with HL Tau, a recently formed star in the constellation of Taurus.

The cause of the disk-like morphology of the proto-Solar Nebula was the rotation of the cloud of gas and dust. A molecular cloud rotates around an axis through it on the same timescale as the Galaxy rotates around its centre. The cloud always shows the same face to the Galactic Centre because the magnetic field of the Galaxy, which threads through the cloud and the other interstellar gas around it, restricts its rotation. Figure 7.2 shows how the magnetic field affects cloud rotation. If the cloud were to rotate so that a somewhat different face were towards the Galactic Centre, the magnetic field in the cloud would be twisted with respect to that in the less dense gas surrounding the cloud. The twist would give rise to a torque which would act to restore the cloud to its original orientation with respect to the Galactic Centre.

If a clump in the cloud were to start to collapse, the magnetic field would restrict that clump's rotation for a considerable fraction of the collapse. However, as the collapse continued, the fractional ionization in the clump and in whatever collapsing fragments that form in it would drop (see Section 5.11). As we described in Section 5.10 a substantial drop in fractional ionization allows relative motions to develop between ions and neutrals, and at very low fractional ionizations the magnetic field does not restrict the flow of the neutral gas. At low fractional ionizations and high neutral number densities, probably in the range 10^{11}–$10^{13}\,\mathrm{m}^{-3}$, the magnetic field is no longer capable of efficiently affecting a clump's rotation or the rotation of a core formed within the clump.

As a core collapses from a number density of between 10^{11} and $10^{13}\,\mathrm{m}^{-3}$, it tends to conserve its angular momentum; i.e. spin-up increases as the core becomes more compact. Everyone is familiar with the example of the spinning skater rotating faster when she pulls her arms inwards; the increase in spin frequency is due to the conservation of angular momentum: all rotating objects that are contracting spin faster as they become more compact. A core with

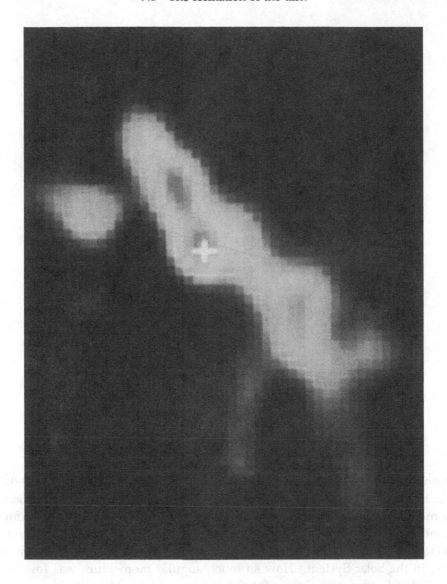

Fig. 7.1. Disk of gas and dust orbiting the star HL Tauri. This is a black and white reproduction of a false-colour map of emission from CO molecules in the gas at a wavelength of 2.6 mm. (From D C Black, *Scientific American*, vol. 240, p.50 (1991 January).)

number density of about 10^{11} m^{-3} is rotating at about the orbital frequency of
the Galaxy, roughly once every 10^8 years, because the magnetic field has
forced it to keep the same face to the Galactic Centre. However, during the
collapse of the core the fractional ionization within it falls and the magnetic
retardation of rotation weakens. Continuing collapse and – in the absence of
outside forces – conservation of angular momentum ensure that the core spin
rate increases.

In reality, not all of the outside forces ever totally vanish, and the angular
momentum is somewhat reduced by them even at high densities, but the

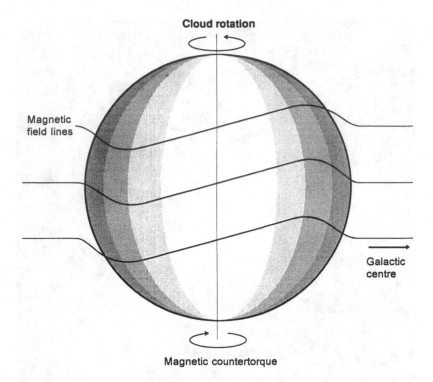

Cloud rotation

Magnetic
field lines

Galactic
centre

Magnetic countertorque

Fig. 7.2. The effect of the magnetic field on cloud rotation. Consider a viewer's frame that orbits the Galactic Centre with the background flow of interstellar medium and stars. Then, if the cloud were to rotate counterclockwise, the magnetic field would be distorted. This distortion produces a torque in the direction opposite to that in which the cloud is rotating and halts the rotation.

spin rate obviously increased from about $10^{-8} \, y^{-1}$ to $1 \, y^{-1}$ for the gas and dust that became the Earth. Even so, the Solar System contains much less angular momentum than a gaseous sphere of the same mass but with a uniform number density of $10^{11} \, m^{-3}$ and rotation period of 10^8 years. (The orbital motion of Jupiter around the Sun carries most of the angular momentum now in the Solar System.) How so much angular momentum was lost from the Solar System remains a fundamental problem. Magnetic forces, the nature of which depended on the chemistry controlling the Solar System's magnetic field structure, probably continued to play an important role in removing angular momentum, but the mechanisms that operated are not understood.

Even though the angular momentum decreased so much, it still affected the Solar System's formation immensely. The orbital velocities of the present day planets are associated with angular momentum which can be thought of as being responsible for the planets maintaining their distances from the Sun. Gravitationally driven collapse proceeded much more easily parallel to the axis around which material was rotating than in other directions, leading to the flattening of the collapsing system and evolution to a disk-like structure.

The quasi-steady disk-like structure that obtained in the proto-Solar Nebula during the era of planet formation can be studied using theoretical models. The model disk is usually taken to possess roughly a solar mass and to survive for a time comparable to the age of the oldest stars observed to have disks, about ten million years. The extent of the model disk is usually assumed to be about the size of the orbit of Pluto. Different assumptions about the friction generated by the rotation of the gas in the disk are made for different models; great uncertainty exists about the way in which friction is produced in turbulent magnetized weakly ionized plasma but the friction is chosen so that most of a disk will lose energy, due to the friction, and fall to the centre in 10 million years. In the following discussion we shall take the typical number density and temperature of the Solar Nebula gas at the orbits of Jupiter, Earth, and Mercury to have been 10^{20}, 10^{21}, and $10^{22}\,\mathrm{m}^{-3}$ and 100, 1000, and 3000 degrees, respectively.

7.2   Lightning in the proto-Solar Nebula?

Much of the information about the proto-Solar Nebula comes from studies of the composition of the meteoritic material which was left over from the era of planet formation. Its solid-state chemical composition was determined by the conditions in the nebula and especially in the outer regions of protoplanets (collections of dust and gas that contracted to form planets). Solid material that became incorporated into planets was heated as a consequence of the release of gravitation energy as the planets formed, and terrestrial rocks have been so heavily processed that their make-up bears no resemblance to that of meteorites. Hence, meteorites are fossils of the young Solar System.

Chondrules, of which many meteorites are partially composed, are small beads of glassy rock having masses of around one hundredth of a gram. Their structures imply that they cooled from temperatures exceeding 1700 kelvins within timescales of tens of minutes to hours. They are believed to have been produced by some type of flash-heating, since those timescales are very short compared to even the smallest possible dynamical times (being a number of years) in the proto-Solar Nebula. Flash-heating could have occurred by several different processes, but one possibility is that lightning struck many regions in the proto-Solar Nebula. For such an electric discharge to occur, the ionization structure must satisfy special conditions. As in other astronomical environments, the ionization structure in the proto-Solar Nebula was controlled by the chemistry.

For lightning to be produced, the electric field must be strong enough to accelerate an electron to a critical energy on a timescale that is short compared to the typical time between the collisions that the electron makes with neutral gas particles. The critical energy is the ionization potential of the neutral species, i.e. the energy required in a collision to knock *another* electron out of the neutral atom or molecule. If the electric field is not strong enough the electron, which loses energy in each collision with a neutral, will not become energetic enough to produce further ionization. Lightning is an avalanche phenomenon that occurs only if the electrons reach high enough energies.

An electric field is produced by the separation of positively charged and negatively charged particles. It may be short-circuited by the motion of electrons and other charged particles, which respond to the electric field and may become distributed to cancel out any charge imbalances. Short-circuiting will be prevented if a large enough fraction of the negatively charged particles and a large enough fraction of the positively charged particles are subjected to forces that are as strong as the electric force and act in opposition to it. For instance, consider a situation in which some heavy positively charged metallic spheres are held by gravity to the Earth's surface, while negatively charged dust particles are suspended by the breeze generated by a fan; furthermore, assume that the metallic spheres and the dust particles are the only important charge carriers. A weak breeze will not suspend the dust particles which will move in response to the electric field (created when they are separated from the spheres) to short-circuit the field. However, if the breeze is sufficiently strong, the charged dust grains will be blown away from the spheres, producing a strong electric field. If an ultraviolet light were allowed to shine on the air between the suspended dust grains, sufficient electrons, which experience much weaker forces from the breeze than do the grains, could be produced so that they could short-circuit the electric field. The ions produced by the ultraviolet radiation could also contribute to the short-circuiting of the electric field because the effect of the breeze on them is also much less than that of the breeze on the dust. Let us use this picture to understand how strong electric fields might have arisen in the proto-Solar Nebula.

Lightning in the proto-Solar Nebula could have been produced if the dominant positive charge carriers were dust grains of one size and the dominant negative charge carriers were dust grains of another size. In this picture, we require the motion of one of these two classes of grains to have been influenced significantly by gravity, while the motion of the other must have been more strongly controlled by the 'winds' that existed in the proto-Solar Nebula. (Such 'winds' were produced by the temperature differences, between the disk's midplane and its outer boundary. Big enough temperature gradients in optically thick media drive convective motions.) Then charge separation and strong electric fields could have been generated, as depicted in Fig. 7.3. If the field strength became great enough a few electrons would have been accelerated to high enough energy to cause further ionization and the onset of the ionization avalanche, i.e. a lightning discharge.

Thus, a necessary condition for the production of lightning in the proto-Solar Nebula is that grains were by far the dominant carriers of both positive and negative charge. In Section 5.11 we described the chemistry that affects the ionization conditions in interstellar clouds. The role of grains in establishing the fractional ionization was also mentioned. At the lower densities that obtain in interstellar clouds, most grains are negatively charged but carry only a small fraction of the total negative charge. However, as mentioned in Section 5.11, the fractional ionization falls with increasing density. At high enough densities the gas phase number density of ions will be lower than the number density of grains;

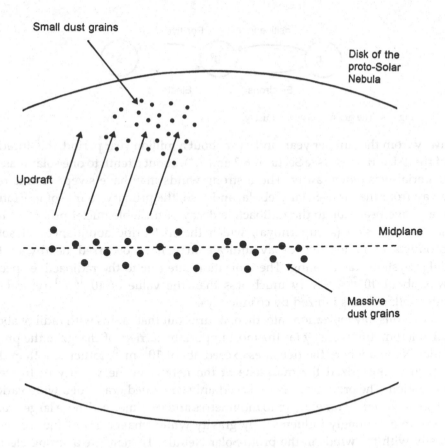

Fig. 7.3. Generation of strong electric fields in the proto-Solar Nebula. Convective (overturning) motions existed in the Nebula because the Nebula was sufficiently optically thick that heat could not be lost rapidly enough by radiation and because the inside was sufficiently hot compared to the outside. Upward motions carried small dust grains upwardly, but large dust grains remained near the disk midplane due to the gravity. If the small grains and large grains carried opposite charges, and the gas phase electron and ion number densities were small enough, large electric fields may have resulted from the separation of the grains of opposite change. If the field became strong enough, an electron avalanche would result.

in such situations, the chemical network controlling the ionization that was described in Section 5.11 must be extended to allow for the charges on the dust grains. Virtually all grains with radii of about 0.1 micrometres in gas with temperatures around 10 kelvins carry charges of −1, or +1, or have zero charge. Grains carrying these charges are easily allowed for in the calculations, and treated as 'normal' chemical species in the ionization chemistry network, consisting of that shown in Fig. 5.12 with the additions depicted in Fig. 7.4.

When calculating the ionization structure in the proto-Solar Nebula, one may reasonably assume that the gas was ionized by cosmic rays at a much lower ionization rate than that appropriate for interstellar clouds. When the Sun first formed it passed through a very high mass loss phase; for perhaps 1000 years it possessed a wind carrying up to one ten thousandth of a solar mass

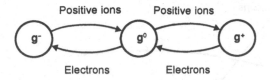

Fig. 7.4. The grain charge chemistry.

away from the Sun per year, and over about a million year period, the duration of the T Tauri phase (see Sections 6.3 and 8.1), about a tenth to one solar mass of material was blown away. These strong winds may have swept cosmic rays away from the proto-Solar Nebula, and if so, the primary source of ionization may have been due to the radioactive decay of unstable nuclei produced in a stellar explosion (a supernova), which the meteoritic abundances of some products of radioactive decay indicate may have occurred nearby as the Solar System was forming. The ionization rate due to the radioactive species was about $10^{-22}\,s^{-1}$, very much less than the value of $10^{-17}\,s^{-1}$ typical for interstellar clouds ionized by cosmic rays.

With this low ionization rate, then, it turns out that grains with radii of about 0.1 micrometres were by far the most important carriers of charge in the proto-Solar Nebula where the densities exceed about $10^{19}\,m^{-3}$, rather less than that which characterized the midplane of the nebula in the vicinity of Jupiter's formation. The proto-Solar Nebula certainly contained grains of a wide variety of sizes with radii ranging up to millimetres and centimetres. These large grains were more strongly influenced by gravity while smaller grains moved more easily with the winds in the proto-Solar Nebula. To generate a strong electric field the large grains and smaller grains must carry average charges of the opposite sign. Electrons probably stuck somewhat more efficiently to large grains than to small grains, since larger grains were built by the collisions of smaller grains and had fluffier surfaces, whereas the more slowly moving ions probably stuck fairly efficiently to all grains. Thus, the larger grains would be expected to have carried a negative average charge while the smaller grains would have been expected to carry a positive average charge. Although this argument is rather speculative it seems likely that the gas phase and grain chemistry controlling the ionization structure in the proto-Solar Nebula created ionization conditions that were favourable for the stimulation of lightning as the planets formed. The calculation of the range of densities, dust grain sizes and fractional abundances, ionization rates, gas densities, and wind speeds required to produce lightning seem to indicate that lightning could have occurred only in very dense regions of the proto-Solar Nebula, those associated with protoplanets.

7.3 Where did the comets form?

A comet possesses a solid head containing various ices. The relative abundances of the molecules frozen into these ices reflect the chemical concentrations in the gas where the comet formed and may be influenced by the melting and evaporation of some of the species during the comet's lifetime. As a comet approaches the Sun, heating of the comet's head melts the outermost layers of ice which then stream off as vapour pushed away from the Sun by the pressure of the solar radiation. As the gaseous material flows through the resulting tail, absorption of the Sun's radiation dissociates the evaporated molecules. Spectroscopic studies, and – in the case of Halley's Comet – direct measurements with a space probe, have led to the inference of the compositions of the ices in comets' heads. In Halley's Comet the ice contains a few times more CO than CH_4, and the N_2 abundance is roughly 1 per cent that of NH_3. It is reasonable to ask: were the CO/CH_4 and N_2/NH_3 abundance ratios anywhere in the proto-Solar Nebula equal to those inferred for Halley's Comet? If the answer were 'yes', potential sites for the formation of comets would have been identified. Hence, we consider relevant aspects of the chemistry in the proto-Solar Nebula.

In regions that are not exposed to significant external sources of radiation or cosmic rays, that have physical properties that do not change over very long times, and that have high enough densities for radiative decays of highly populated excited levels to occur at insignificant rates compared to the rates at which particles collide, the chemistry attains a special form; it is said to be in *thermodynamical chemical equilibrium*. It obtains in the hotter parts of the proto-Solar Nebula and in the most interior regions of some of the stellar outflows to be described in Chapter 8. It does *not* obtain in any of the environments that have been treated in Chapters 4–6. In those chapters, we often considered sources in which the timescales associated with significant changes in physical conditions were short compared to the timescales on which chemical reactions occurred; for instance, once it begins the collapse of one of the initial state clumps described in Section 6.3 probably occurs faster than many atomic species can be removed as molecules are created. However, even in many objects in which physical conditions vary slowly the effects of outside radiation or cosmic rays prevent thermodynamical chemical equilibrium from being reached. In other very stable sources the density is too low to allow the attainment of thermodynamical chemical equilibrium. However, when thermodynamical chemical equilibrium conditions are met, the chemical abundances depend simply on the temperature and density. The abundances can be calculated straightforwardly from a knowledge of the energy level structures of the molecular species, obtained using laboratory spectroscopic techniques.

Some results for the abundance ratios of N_2/NH_3, CO/CO_2 and CO/CH_4 calculated using a thermodynamical chemical equilibrium model are given in Fig. 7.5. The model results are probably valid only for relatively high-temperature gas that was roughly as near or nearer to the Sun than the Earth presently is. The basic assumption (that reaction sequences occurred on timescales short compared to the 3–10 million years that the proto-Solar Nebula existed) on which the model is based was most likely violated in

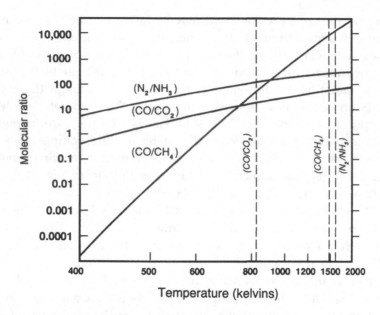

Fig. 7.5. Proto-Solar Nebula abundance ratios of several species for a thermodynamical chemical equilibrium model. The model abundance ratios of N_2/NH_3, CO/CO_2,
and CO/CH_4 are given as functions of temperature for an assumed relationship
between the density and temperature in the proto-Solar Nebula. If the reactions
that dominate the chemistry all occur in the gas phase and no grain surface catalysis
is important, the model results for a particular ratio are valid only at temperatures
exceeding that marked by the corresponding line. (From B Fegley and R G Prinn
in *The Formation and Evolution of Planetary Systems*, eds. H A Weaver and L Danly,
Cambridge University Press, Cambridge (1989).)

lower-temperature parts of the Nebula. Possibly, grain catalysis was important
in the production of gas phase species, speeded up the chemical sequences, and
resulted in the model being valid at lower temperatures than might otherwise
be expected.

In any case, many meteoritic remnants (but not chondrules) have physical
structures that are consistent with them having been heated and then slowly
cooled, perhaps several times. Repeated heating and cooling may have arisen
due to the random transport of grains towards and then away from the
proto-Sun by the turbulent eddies that were present in the proto-Solar
Nebula. The eddies may have caused significant mixing of gas, in addition to
the mixing of grains that became parts of meteorites, from the nebula's hotter
inner regions, where thermodynamical chemical equilibrium obtained, out
into the cold regions. Thus, the chemical composition in the cold regions,
where reactions were too slow to alter the make-up of the gas transported
out from the hotter regions, may have been similar to that nearer to the Sun.
Very efficient mixing would have made the chemical composition of much of
the proto-Solar Nebula appear 'hot'.

The results in Fig. 7.5 show an N_2/NH_3 ratio that is more than several
thousand times that inferred for Halley's Comet ices, while the regions in the

nebula where the temperatures were about 700 kelvins had a Halley-like CO/ CH_4 ratio if thermodynamical chemical equilibrium obtained. If the cometary N_2/NH_3 ratio is identical now to what it was when the comet formed, the N_2/NH_3 ratio in Halley's Comet clearly cannot be explained by the formation in a proto-Solar Nebula in which thermodynamical chemical equilibrium existed or in which the mixing, discussed in the previous paragraph, between inner and outer regions of the nebula made the whole nebula possess a 'hot'-inner-type chemical composition.

The Nebula abundances may have been altered by mixing of gas from embedded inhomogeneities, if the mixing took place on timescales faster than the chemical timescales. The proto-planetary condensations from which Jupiter and Saturn formed may have provided sources of chemicals that mixed into the Nebula gas. The proto-planetary condensations were denser than the surrounding background Nebula. If thermodynamical chemical equilibrium applies then high densities result in high fractional abundances of hydrogen-bearing species like CH_4 and NH_3 if the dominant element was hydrogen, as it was in the proto-planetary condensations from which the largest planets formed. (As they collapsed, the condensations which became the smaller planets such as Earth lost most of the hydrogen because their gravitational fields were too weak to prevent molecular hydrogen from boiling off, though heavy grains continued to fall.) Perhaps, comets formed in the vicinities of the proto-Jupiter and proto-Saturn condensations. Proto-planetary material with a low N_2-NH_3 abundance ratio may have mixed with background proto-Solar Nebula material with a $CO-CH_4$ abundance ratio greater than 1.

Another possibility is that the comets formed in a part of the proto-Solar Nebula that was too cold and too poorly mixed for chemical abundances to be altered from the values that they had in the interstellar gas before it collapsed. Theoretical models of the chemistry in collapsing dark cores in which stars form include the freeze-out of gas phase species onto grains (see Section 6.5). It is possible with such a model to calculate the composition of the ices that would exist on the grains by the time that most species containing atoms heavier than helium have almost totally frozen out onto dust grains. It is intriguing that one model designed to give chemical abundances similar to those in a well-observed dark core predicts relative abundances of CH_4, H_2O, and NH_3 ices that agree very well with those measured for Halley's Comet. This agreement may imply that Halley's Comet formed far out in the proto-Solar Nebula from material that had such a quiescent dynamical history that the ice content left over from the era of the onset of dark core collapse was relatively undisturbed. This picture is very different from the one in which warm proto-Solar material somehow mixed dynamically with warm proto-Jovian and/or proto-Saturnian material to form comets. The collapsing dark core model for the chemistry and for ice production gives far higher ratios of CO/CH_4 and of N_2/NH_3 than are observed in Halley's Comet; the relatively low abundances in Halley's Comet of CO and N_2 ices have been attributed to the rapid ways in which CO and N_2 ices diffuse through other ices (e.g. CH_4, H_2O, and NH_3) and evaporate. It is likely that Halley's Comet has lost far larger fractions of its initial CO and N_2 than its initial CH_4, H_2O, and NH_3 ice components, and a reasonable goal is to explain only the measured CH_4, H_2O, and NH_3 ice abundance ratios.

7.4   Why does the Earth have water?

The Earth possesses many features that make it far more suitable for life as we know it than the other planets in the Solar System. Each of these features is the consequence of the interaction of so many processes that theoreticians at present would fail to predict them if working in the absence of empirical knowledge. As an example of these features that arose due to the nature of the proto-Solar Nebula chemistry, we consider the existence of Earth's water.

The proto-Earth did not have a sufficiently strong gravitational field to retain proto-Solar Nebula gas during planet formation. Rather, the Earth is composed of material contained in grains and the larger stony structures that were composed by the grains agglomerating together. The Earth's atmosphere and surface water have been released from the stony interior primarily by volcanic activity.

Venus has only about a hundred thousandth as much water in its atmosphere and on its surface as the Earth. It has lost a greater percentage of the water that has been outgassed than the Earth has. (This can be inferred by comparing the Venus HDO/H_2O ratio with the terrestrial HDO/H_2O ratio. HDO, being heavier than H_2O, is retained by a planet more easily than H_2O; hence, the HDO/H_2O ratio increases in a well-understood way as H_2O is lost. The primary loss mechanism for H_2O may be atmospheric ablation by the solar wind.) Still, the Venusian rock probably contained only about 0.3 per cent as much water as the Earth rock did at the time of planet formation.

The difference in the initial water contents of the rocks of the two planets is unexplained. Thermodynamical chemical equilibrium models show that many hydroxyl rich minerals, like serpentine $(Mg_3Si_2O_5(OH)_4)$ and talc $(Mg_3Si_4O_{10}(OH)_2)$, from which water can be formed during planetary outgassing, can be abundant only in fairly low-temperature regions of the proto-Solar Nebula. However, in those regions the reactions that form such minerals are too slow to do so during the lifetime of the Nebula, and, hence, the models greatly overestimate the abundances of those species. There is some possibility that CO clathrate, $CO.H_2O$, which could be retained in stony material was abundant in the low-temperature regions of the nebula, as suggested by thermodynamical chemical equilibrium calculations, but the rates of the reactions that formed it are uncertain.

Possible sources of minerals from which the Earth's water came are the higher density proto-Jupiter and proto-Saturn environments. Mixing of these minerals into the proto-Solar Nebula near them and then throughout the proto-Solar Nebula would have then been necessary. If Venus formed in a sufficiently high-temperature region some of these minerals would have been destroyed, leading to its potential source of outgassed water being smaller than Earth's.

Another possibility is that late in the formation of the Earth it accreted grains with icy mantles. Such mantles might have been destroyed partially in the higher-temperature region of the proto-Solar Nebula where Venus was born. The icy mantles may have been remnants of the interstellar chemistry that occurred as the Nebula collapsed.

The various hypotheses for the origin of the Earth's water must be in harmony with the abundance of water on moons of the giant planets, the mineral contents of meteorites, and the low water abundances of other planets.

7.5 The Sun as a molecular source

We now consider an astronomical object which belongs to the Solar System and which has been a molecular source from the birth of the Solar System up to the present.

Most of the Sun's surface at about 6000 kelvins is too hot for molecules to survive there. However, the Sun's outer layers do contain cooler regions called sunspots. Each sunspot consists of gas that is thermally insulated by the local configuration of the magnetic field from the hotter gas around it.

The energy generated by nuclear reactions in the Sun's interior is transported to gas just below the surface by a combination of the propagation of radiation emitted by the hot gas and of convective motions of hot gas rising from the interior, releasing its heat, and sinking back towards the centre to be reheated. However, much of the energy deposited somewhat below the surface of the Sun is transported upwards by thermal conduction, a mechanism familiar to anyone who has held one end of a piece of metal while placing the other end in a flame. Efficient thermal conduction in a highly ionized gas requires the uninhibited diffusion of the hotter ions and electrons towards cold regions as cold ions and electrons diffuse towards warmer regions. A magnetic field suppresses this diffusion in directions perpendicular to the field direction. Thus, thermal conduction is inefficient in these directions. Some parcels of gas at the solar surface are wrapped in a magnetic field and are insulated thermally. They cool to around 4000–4500 kelvins, and, though they are emitting radiation they emit it at lower rates than the hotter parts of the Sun's outer envelope and appear to be dark sunspots. Above a sunspot the temperature drops to about 3200 kelvins over a small region of space.

Gas at 3200 kelvins is cool enough for the formation of H_2 by the same sequences that produced it in the era of galaxy formation (see Chapter 4) and which play a role in the cool outer envelopes of some stars (see Chapter 8). Ultraviolet emission from solar H_2 has been detected, making the Sun the largest molecular source in the Solar System.

Selected references

Fegley, B Jr and Prinn, R G: 'Solar Nebula Chemistry: Implications for Volatiles in the Solar System', in *The Formation and Evolution of Planetary Systems*, eds. H A Weaver and L Danly, Cambridge University Press, Cambridge (1989).

Jordan, C, Brueckner, G E, Bartoe, J-D F, Sandlin, G D and van Hoosier, M E: 'Lines of H_2 in extreme ultraviolet solar spectra', *Nature*, vol. 270, p.326 (1977).

Nejad, L A M, Hartquist, T W and Williams, D A: 'Models of dense cores in translucent regions of low-mass star formation', *Astrophysics and Space Science*, vol. 220, p.261 (1994).

Pilipp, W, Hartquist, T W and Morfill, G E: 'Large electric fields in acoustic waves and the stimulation of lightning discharges', *Astrophysical Journal*, vol. 387, p.364 (1992).

Stellar winds and outflows

———

An', as it blowed and blowed, I often looked up at the sky an' assed meself
the question – what is the stars, what is the stars?

Sean O'Casey in *Juno and the Paycock*, Act 1

As indicated already in Chapters 6 and 7, during the first million years or so of
the life of a star with a mass comparable to that of the Sun the star loses a good
fraction of a solar mass of material in a wind. As this period of high mass loss
ends the star settles down into the relatively quiescent period of what is called
its 'main sequence' phase of evolution.

Like many chemical reactions, *nuclear* reactions have energy barriers; these barriers are due to the electrostatic repulsion between nuclei, and high temperatures (of the order of tens of millions of degrees and more) are required for nuclei to overcome the electrostatic barriers between them and approach each other closely enough (10^{-14} m separation) to interact via the strong nuclear force. Hence, nuclear burning occurs only in the hot central regions of the Sun and not in the cooler outer envelope. A star remains in its main sequence phase until it burns most of the hydrogen in its hot central region to form helium. At the end of the main sequence phase the star has a core composed mostly of helium.

Helium can also burn through nuclear reactions, and the product of helium burning is carbon because the primary helium burning reaction involves the simultaneous collision of three helium nuclei which form a complex which is stabilized by the loss of energy to form the carbon. The electrostatic repulsion between helium nuclei is greater than that between hydrogen nuclei, and helium burning takes place at higher temperatures than hydrogen burning. These higher temperatures are initially reached in the stellar core by the gravitationally induced contraction of the core, since rapid compression of gas results in its heating. Then the helium burning begins. A thin shell of hydrogen also burns just outside the helium core; this hydrogen was too cold to burn during the main sequence phase of the evolution but became hotter during the helium burning phase initially in part because of the heating due to contraction and in part because of the energy release during the helium burning. Though higher temperatures must be established in the core of a star like the Sun to start this phase of its evolution, once those higher temperatures are established the nuclear burning is much more rapid during the helium core–hydrogen shell burning phase of the star's life than during its main sequence phase. The faster release of energy results in a higher rate of energy input into the outer envelope of the star. The outer envelope expands, and a star with a mass similar to that of the Sun will enter a phase during which its outer envelope expands to a radius approaching 100 times the radius of the Sun.

As sufficient carbon is produced in the stellar core, a predominantly carbon core surrounded by a helium burning shell which is in turn surrounded by a hydrogen burning shell comes into existence. Carbon reacts with helium will release energy and form oxygen, but these reactions occur only after even further core collapse takes place since even higher temperatures are required to drive the helium burning reactions. During this phase in which a carbon burning core is surrounded by a helium burning shell encased in a hydrogen burning shell, the rate at which energy is deposited into the outer envelope of the star is so great that the envelope expands away from the star and is no longer bound by gravity to it.

Much of this chapter is concerned with mass loss during the earliest period of the life of a star (the so-called T Tauri phase), and mass loss during the carbon burning core–helium burning shell–hydrogen burning shell phase of

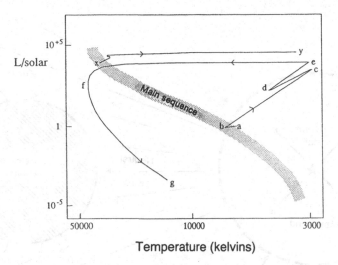

Fig. 8.1. A diagram plotting luminosity versus surface temperature, showing approximate evolutionary tracks for stars of two different initial masses. Path a–g is for a star of about 1 solar mass. Path x–y is for a star of about 10 solar masses. Stellar temperatures are measured in kelvins.

At a, the solar mass star, having recently formed in an interstellar cloud, is passing through a T Tauri phase (see Section 8.1), before joining the main sequence at b, where it remains for about 10^{10} years. The star then expands and cools, becoming a red giant at c; then warms up and fades a little, joining the horizontal branch at d, before going through a second red giant phase, on the so called 'asymptotic giant branch' (AGB) at e, during which heavy mass loss occurs as a 'superwind' (see Section 8.2). Exhaustion of the AGB atmosphere exposes the hot luminous stellar core, which moves to point f. Ultraviolet radiation from this core (now only about 0.6 solar masses) ionizes the inner part of the remnant superwind, forming a planetary nebula (see Section 8.3). Dispersal of the remnant superwind, and fading of the core star leave it as a white dwarf, at position g. A nova (see Section 8.4) occurs in some binary systems, consisting of a red giant (c) plus a white dwarf (g), where mass transferred to the latter builds up to the point where the accreted matter suddenly initiates nuclear fusion. The 10 solar mass star leaves its position on the main sequence, at x, becoming a supergiant at y, where it explodes as a supernova (see Chapter 10).

a low-mass star's life. By low-mass star we mean one with a mass less than about six times that of the Sun. The evolutionary picture that we have described is appropriate for such stars and results in the births of stellar remnants called white dwarfs. We consider the fate of a more massive star in Chapter 10. The evolution of low- and high-mass stars is depicted in a luminosity–temperature plot (Fig. 8.1).

Molecular chemistry plays important roles as a probe of the mass loss of low-mass stars and, in some cases, as an agent which modifies the physical nature of the mass loss. We consider these two roles of molecules in this chapter as well as the production of some molecules whose presence in the sources is of interest in their own stellar environments or helps substantially in the diagnosis of them.

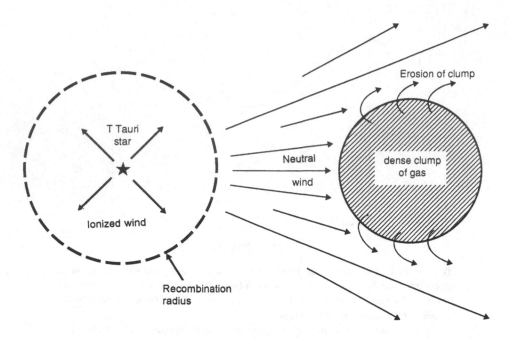

Fig. 8.2. The wind from a T Tauri star and its interaction with a dense clump of gas (not to scale).

8.1   T Tauri winds

We first focus on the winds of very young, fairly low-mass stars, belonging to a class designated by the name of the star that typifies this family, T Tauri. The structure of a T Tauri wind is depicted in Fig. 8.2. As we saw in Chapter 6, low-mass stars form in dense cores within molecular clouds. Their winds have a dramatic effect on those cores, eroding and dissipating them entirely. Indeed, the winds drive the dynamics of the entire molecular cloud. Although the molecular flows driven by the stellar winds can be detected, these flows are not the stellar winds themselves but molecular clump material entrained by the wind. The nature of the interaction between the stellar wind and the molecular cloud is not well understood, nor is the actual nature of the wind. If we could observe molecules in the wind itself then we would be able to understand much better the nature of that primary driving force. So the question we ask is this: is a substantial chemistry likely to develop in this wind? If so, for which molecules should we search?

The wind of a T Tauri star has its origin in energetic motions inside the star. These motions generate waves which continuously heat and accelerate the tenuous outer layers of the star's atmosphere, creating a wind which removes mass from the star at a rate estimated from observations of the winds to be typically about 10^{-7} solar masses per year. This wind is expected to reach a peak temperature of several thousand kelvins at a distance of about three stellar radii from the centre of the star, at which point the typical number density in the wind is about $10^{15}\,\mathrm{m}^{-3}$. (The radius of a T Tauri star is about $2 \times 10^9\,\mathrm{m}$, somewhat larger than that of the Sun which has a radius of $7 \times 10^8\,\mathrm{m}$.) Under these conditions, collisions between atoms, ions, and electrons in the gas are sufficiently energetic and frequent that the gas is almost completely ionized. The ionization is caused solely by collisions; a T Tauri star emits very little ultraviolet radiation that is capable of ionizing hydrogen. The gas is also dust-free; even if dust grains were present they would be rapidly eroded under these conditions.

As the wind flows outwards, the gas expands and radiates, so that its temperature and density both fall. As a result, ionizing collisions are less effective and less frequent, while the recombination of ions and electrons occurs more readily at lower temperatures. Consequently, there is a point in the flow at which the gas becomes mainly neutral rather than ionized. The radius at which this happens, the recombination radius, is typically about six times the stellar radius. The recombination in the expanding flow is never total, so there is residual ionization outside the recombination radius. The question we ask is this: can molecules form in this outflow? The wind effectively rejoins the interstellar gas when it has travelled about a thousand radii from the star, a distance that a parcel of gas covers in under a year. Is there time for chemistry to occur?

This situation is unlike any other we have so far considered. The gas is warm, mostly neutral, and dust-free, and in these respects is similar to gas in the Early Universe. But in T Tauri winds the important elements oxygen, carbon, and nitrogen (and others, too) are present. A chemistry based on H_2 may be fast, but the difficulty is to make enough H_2. The routes adopted in our discussion of chemistry in the Early Universe (Chapter 4) involved electrons and protons as catalysts. Specifically, electrons reacted with H atoms to form H^- and emit radiation; H^- then collided with H to produce H_2 and a free electron. Also, H^+ reacted with H to form H_2^+ and emit radiation; H_2^+ then reacted with H to make H_2 and H^+. In the present situation, however, the T Tauri star emits enough infrared radiation so that the reaction

$$H^- + \text{infrared radiation} \rightarrow H + e^-$$

maintains H^- at a low enough abundance for the second sequence of reactions, involving protons, to be more important.

When the temperature is high, many reactions with H_2 are possible (see Section 5.8) and a chemistry develops that forms CO quite readily. O and C react with H_2 to form OH and CH. Then OH reacts with C and CH with O to form CO. As CO is not easily destroyed it accumulates. However, once the gas has cooled then the reactions of O and C with H_2 are suppressed, and the chemistry proceeds only by slow radiative processes in which only about one in every 10^5 collisions usually leads to reaction. Theoretical studies of this situation indicate that molecules such as CO are unlikely to be produced in detectable amounts unless the density in the flow is sufficiently high, requiring the mass loss rate of the star to be considerably larger than 10^{-7} solar masses per year. Detection of CO in the original stellar wind would, therefore, help to determine the mass loss rate of the star, which can be estimated from other observational data only very roughly.

If the wind is eroding a dense clump of gas left over from the star formation process, then the wind gas can be enriched with molecular hydrogen. If this occurs, then much of the available carbon can be converted to CO, and these amounts would certainly be detectable.

8.2   Cool stellar envelopes

Very rich chemistries occur in the extended envelopes around some highly evolved stars having masses up to a few solar masses. These are stars that are shedding their outer envelopes during the phase in which their structure is carbon burning core–helium burning shell–hydrogen burning shell. This is particularly evident in the case of envelopes that are carbon-rich. In one such envelope the exotic molecules HCN, HC_3N, HC_5N, ..., $HC_{11}N$ have all been detected, together with a wide variety of more conventional species. In this section much of our discussion concerns molecules whose existences we deem to be interesting in their own right but which play little dynamical or diagnostic role. However, near the end of the section we stress important diagnostic and dynamical functions of the chemistry in such envelopes.

The ejected envelope of a low-mass star contains about one tenth to several solar masses and expands with a speed of 10–$20\,km\,s^{-1}$ for a period of perhaps $10\,000$ years, so that its extent may be about a light year. Eventually, the central core of the star shrinks, becomes very hot, reaching a temperature of several tens of thousands of kelvins, and emits ultraviolet radiation, ionizing the gas around it and creating a new type of source, the nature of which we shall describe in the next section.

But while the star was cool, conditions in its atmosphere were ideal for a special kind of chemistry. At a radius of about 4×10^{10} m the gas temperature was about 2000 kelvins and the number density in the gas was about 10^{19} H_2 molecules m^{-3}. These conditions were easily sufficient to allow three-body collisions to occur. In these situations the chemistry differs greatly from that which we have normally considered so far. The various atoms are able to interchange between the molecules many times until they have arranged themselves in the most stable forms possible at that temperature and density. For example, the atoms C and O are generally more strongly bound in CO than in CH and OH. If oxygen is more abundant than carbon, then the excess oxygen can appear as H_2O, CO_2, NO, etc. This is called the chemistry of thermodynamic equilibrium (see Section 7.3) and can be calculated in a straightforward way, if the bond energies in all the molecules are known. For a carbon-rich environment, the abundant molecules are likely to be CO, N_2, C_2H_2, CH_4, NH_3, and HCN, though the precise composition depends on the elemental abundances, the density, and the temperature. In addition, dust particles may form. In a carbon-rich gas, particles of soot may appear, as

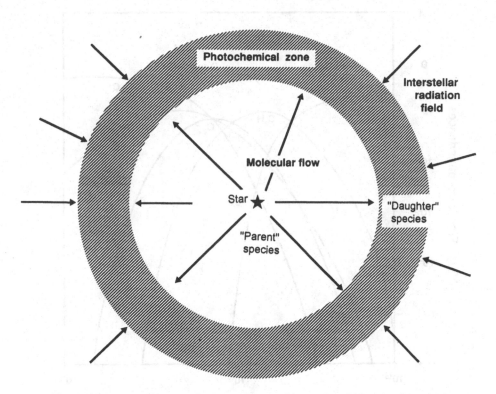

Fig. 8.3. Schematic diagram showing location of 'parent' and 'daughter' species in circumstellar envelopes.

they do in a candle flame. If the gas is oxygen-rich then silicates may become supersaturated and may condense out as a solid grit. Evolved stars with carbon-rich envelopes exist, as do those with oxygen-rich envelopes.

This mixture of molecules, and – possibly – dust, is how the gas starts its journey to interstellar space, as is depicted in Fig. 8.3. As the gas flows away from a star, the density falls and the temperature also falls. The frequency of collisions between atoms and molecules and the energy of those collisions also fall. There comes a point when the atoms cannot easily interchange between the various molecules, and the relative molecular composition is, therefore, fixed at that point. The relative abundances of the molecules are 'frozen into the gas' as the gas flows outwards, and in these relative abundances the gas carries with it a 'memory' of the time when it was warmer and denser.

However, this situation does not persist indefinitely. Although most of the stars that have extended envelopes are so cool that they do not generate an ultraviolet radiation field, the radiation field from interstellar space penetrates into the dusty envelope, so that as a parcel of gas moves away from the star, the radiation field gets stronger and stronger since the material outside of it shielding it from the galactic background radiation becomes more and more diffuse. The molecular dissociations and ionizations, in particular, caused by

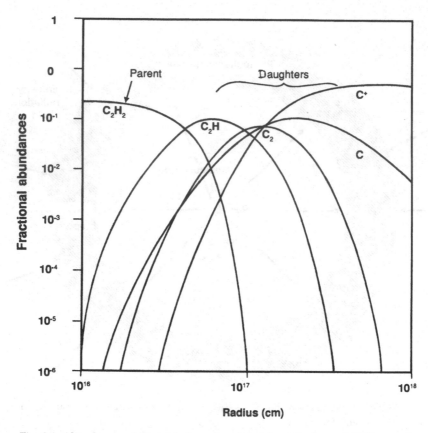

Fig. 8.4. Abundances of 'daughter' species arising from the 'parent' C_2H_2, varying with radius from the star. (Adapted from L A M Nejad, T J Millar, and A Freeman *Astronomy and Astrophysics*, vol. 134, p.129 (1984)).

the ultraviolet radiation become more frequent. (Note that some of the stars with oxygen-rich envelopes do have weak ultraviolet fields that modify but do not totally suppress the chemistry; see below.) For example, in the atmosphere of a carbon-rich star one of the important carbon-bearing molecules formed in thermodynamical chemical equilibrium is acetylene, C_2H_2. The radiation field converts this to a succession of products including C_2H, C_2, C, and finally C^+ and in the absence of any chemical reactions in the outflow we would expect to find these species at successively greater distances from the star, as depicted in Fig. 8.4. Similarly, ammonia, NH_3, in the stellar atmosphere is photodissociated in the outflow to give NH_2, NH, and N, and HCN is photodissociated to give first CN, then C and N, and finally C^+ and N. Ultimately, as the flow merges into the diffuse interstellar medium, it is composed mainly of atoms H, O, N, and the ion C^+.

However, the new species generated in the outflow from the 'parent' species C_2H_2, NH_3, HCN, such as C_2H, C, C^+, NH, CN etc, are themselves reactive. There is a *photochemical zone* between the region where the 'parent' molecules are formed and the region in which the gas contains only atoms and atomic ions. In the photochemical zone the 'parent' molecules mix with the reactive products of their destruction, and there is the opportunity for reaction to occur between these new species and parents, or between the new species. For example, C^+ ions react with molecules in a variety of ways:

$$C_2H_2 \xrightarrow{C^+} C_3H^+ \xrightarrow{H_2} C_3H_3^+ \xrightarrow{e^-} C_3H_2$$

initiating with acetylene, its parent, a path which leads to a new species in the outflow, C_3H_2, cyclopropylyne, a cyclic molecule with the unusual structure

$$HC \equiv\!\!\!\!\overset{\displaystyle C}{\diagup\diagdown}\!\!\!\! CH.$$

Cyclopropylyne has also been detected in interstellar space and appears to be very common.

The 'daughter' species C_2H is a progenitor of larger hydrocarbons

$$C_2H \xrightarrow{C_2H_2} C_4H_2 \xrightarrow{radiation} C_4H$$

a linear hydrocarbon radical that is also detected in a nearby star with a massive carbon-rich outflow called IRC + 10216. The cyanopolyyne sequence, HCN, HC_3N, HC_5N, ..., $HC_{11}N$, detected in IRC + 10216, is one of the most surprising features of circumstellar chemistry and astrochemistry in general. Some of these molecules were actually detected in space before they had been synthesized in the laboratory. It is still not clear how they are formed in space. The simplest molecule in the sequence, HCN, is probably a 'parent' molecule. It seems possible that the 'daughter' ion, $C_2H_2^+$, created by photo-ionization of the 'parent' C_2H_2, may take part in reactions that increase the complexity.

Once the photodissociation and photoionization of parents begins, many chemical pathways are opened. However, the complexity of the products that can be created is limited by the amount of time that any parcel of gas spends in this reactive zone. The molecules formed in this zone experience a stronger and stronger radiation field until the field intensity is almost equal to that of the unshielded interstellar medium. Increasing the density in the flow helps the chemistry, so stars with higher mass loss rates should have richer chemistries. Until now much of the focus of the study of chemistry in the photochemical zone has been concerned with understanding the formation of the relatively complex species such as C_3H_2 and HC_5N, but in the future detailed studies of the chemistry may be used to determine not only the mass loss rate of the star, but also the relative abundance of the elements in the gas, the local radiation field intensity, and other important and useful astronomical parameters.

When an oxygen-rich envelope develops around a cool star, the chemistry will be much more restricted than in a carbon-rich envelope, such as that around IRC+10216. In an oxygen-rich environment, nearly all the carbon will be in carbon monoxide, CO, and very little is available to feed the extensive hydrocarbon chemistry that a carbon-rich star exhibits. Some oxygen-rich stars are observed to lose mass at a rate comparable to the carbon-rich stars, but have sources of ultraviolet radiation arising in the stellar chromosphere. This radiation modifies the chemistry extensively so that few molecules are observed. Nevertheless, it is in such envelopes as these that intense emission from silicon monoxide, SiO, is sometimes observed. It arises in SiO masers; we shall discuss them in Chapter 9.

So far in this section we have described chemistry that until now seems to serve neither a diagnostic function nor to play a major part in the evolution of the envelopes. However, there is a chemistry, albeit a poorly understood one, that has major consequence for the outflows. This chemistry is the one that initiates the formation of dust observed by its infrared emission to exist in the envelopes. The light from a star exerts a pressure on the dust in its vicinity. This so-called *radiation pressure* accelerates the dust which drags the gas along with it: the velocity of the envelope as it moves away from the star is determined by the interaction between the stellar light and the dust formed in the envelope as a consequence of the chemistry. Some of the molecules produced en route to the formation of dust act as coolants allowing the temperature in the envelope to drop enough that solid dust grains can condense.

8.3   Planetary nebulae

Eventually the outer envelope of a star with a mass of up to several solar masses is lost, and the remaining central parts of the star undergo contraction. A surface temperature of many tens of thousands of kelvins (as compared to the Sun's surface temperature of about 6000 kelvins) is reached. The star's evolution at this stage is not entirely understood but for a timescale of tens of thousands of years the star emits tens of thousands of times as much energy as the Sun. Eventually, however, the star cools to several thousand kelvins and dims to about one ten thousandth the luminosity of the Sun. By then it is a white dwarf with a radius of only a few per cent that of the Sun.

During the hot phase of the star's evolution it develops a wind which carries about 10^{-7} solar masses away from the star each year at a speed of several thousand kilometres per second. Just as the evolution of the star's radiation field is highly uncertain, the characteristics of the development of this fast wind are obscure. However, this phase of stellar evolution is a very interesting one, and astronomers would like to understand much more about how a star so spectacularly enters its dotage.

The high emission rate of ultraviolet photons associated with the heating of the stellar surface and the fast wind have significant dynamical consequences for the surrounding, recently ejected envelope of the progenitor star. In fact, the radiation and wind interactions with the envelope create in about ten thousand years one of the most beautiful of astronomical phenomena, a planetary nebula, an example of which is pictured in Fig. 8.5.

The name planetary nebula was given to these objects because, like planets, they appeared as small disks of light in early small telescopes, whereas stars are

Fig. 8.5. The Helix Nebula, NGC 7293 (copyright 1981 AAT Board).

point-like sources. The envelope of roughly 1 solar mass of material extended over about a light year at a number density of about $10^9 \, m^{-3}$ in a planetary nebula is ionized and heated to around 10^4 kelvins by the stellar radiation. Though one might assume that a planetary nebula is a hostile environment for molecules, molecular emissions from some planetary nebulae are observed. A number of possible scenarios for molecular formation have been considered; perhaps, as described below, studies of molecules in planetary nebulae will not only lead to the identification of which scenario is correct but also give useful insight into the evolution of low- and moderate-mass stars after they have shed their outer envelopes.

In Section 4.3 we considered the way in which a wind interacts with an external medium. To understand planetary nebula formation we must consider the simultaneous effects of a wind and a radiation field on the external medium. Figure 8.6 shows the structure of a radiation-modified wind-blown bubble propagating into an envelope. The wind drives a shock through the envelope, which is clumpy as is indicated by the observational detection of clumps in planetary nebulae. For typical wind and envelope conditions the shock moves at only around $10\text{--}20 \, km \, s^{-1}$ relative to the envelope gas which itself is moving outwards at $10\text{--}20 \, km \, s^{-1}$ or so. However, the shock does cause heating to a couple of thousand kelvins. Because the clumps are dense (they are at a lower temperature than the rest of the envelope though they are at roughly the same pressure) the shock propagates more slowly into them and probably fails to heat them to the temperatures required to drive the endothermic reactions that might occur in the rest of the shocked envelope gas. Clumps are able to survive even in the parts of the nebula where the wind fills most of the volume. The boundary of the ionized material inside the shock moving into the envelope is shown in Fig. 8.6. This boundary is in the shocked envelope gas and moves outwards as the shock does.

In fact, when drawing Fig. 8.6 we *assumed* that the boundary of the ionized gas is closer to the star than the shock propagating outwards through the envelope. In reality, too little is understood about the history of the increase in stellar ultraviolet emission and the development of the fast wind to know where this shock is with respect to the ionization boundary. Possibly, radiation ionizes all of the envelope, except the clumps which are more opaque, before the envelope is shocked. Of course, the relative locations of the shock and ionization boundary have tremendous importance for the possibility of molecular formation. This means that the molecular emissions are potentially useful for diagnosing the nebular dynamics and, hence, the evolution of the central star.

For the rest of this section we shall assume that the relative positions of the ionization boundary and the shock in the envelope are qualitatively similar to those in Fig. 8.6. Then there are potentially three different types of molecular region. They are: the unshocked envelope gas; the clumps; and the shocked but not yet ionized envelope gas.

In the first two types of region molecules observed now may simply be remnants of the chemistry that occurred in the period when the envelope was much closer to the star and three-body reactions were important (see the previous section). Of course, photons may be processing these chemical remnants or possibly even releasing molecules into the gas phase from grains by heating them. The photoprocessing is likely to be particularly effective in the clump-free parts of the envelope and (depending on how the stellar radiation field evolves) may reduce the clump-free, unshocked gas to atoms on a timescale of about a hundred years as compared to the dynamical timescale of about 20 thousand years associated with a mature planetary nebula. The clumps are more opaque, and the lifetimes of molecules against photodestruction may be tens of times longer than in the clump-free unshocked envelope gas. The idea that chemicals may be molecular remnants gains some support from observations of emissions from the cyanopolyynes HC_3N and HC_5N from planetary nebulae that appear to be very young. These molecules are abundant in the envelopes of carbon-rich stars when the envelopes are still closely associated with them and before the radiation field evolves. A careful study of the dynamics of these young planetary nebulae and the cyanopolyyne emissions may yield useful information about the growth of the stellar luminosity as the envelope is lost.

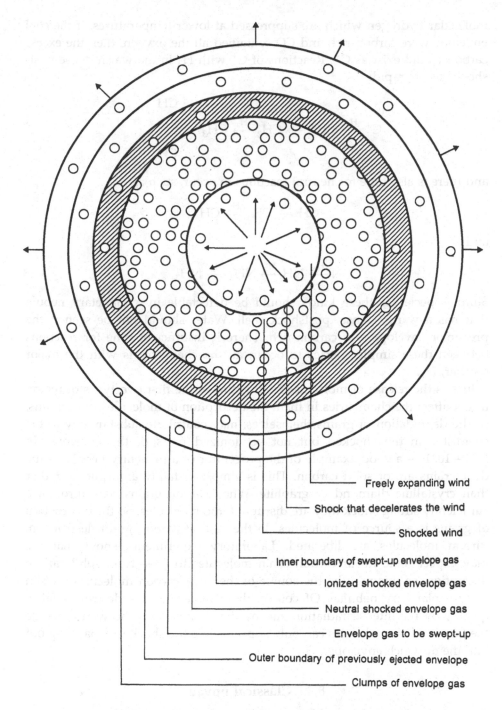

Freely expanding wind

Shock that decelerates the wind

Shocked wind

Inner boundary of swept-up envelope gas

Ionized shocked envelope gas

Neutral shocked envelope gas

Envelope gas to be swept-up

Outer boundary of previously ejected envelope

Clumps of envelope gas

Fig. 8.6. The structure of a developing planetary nebula.

The third possible type of region where molecular abundances may be significant is the nonionized but shocked portion of the envelope (if the shock actually precedes the ionization boundary). In the dense, warm, post-shock shell conditions are ideal for driving reactions between atoms and molecular hydrogen which are suppressed at lower temperatures. If the cool envelope were carbon-rich and CO contained all the oxygen, then the excess carbon would exist as C^+. Reactions of C^+ with H_2 in the warm dense post-shock gas are rapid:

$$C^+ \xrightarrow{H_2} CH^+ \xrightarrow{H_2} CH_2^+ \xrightarrow{H_2} CH_3^+ \begin{array}{c} \xrightarrow{e^-} CH \\ \xrightarrow{e^-} CH_2 \end{array}$$

and there is also time for neutral reactions to occur, such as

$$CH \xrightarrow{H_2} CH_3 \xrightarrow{H_2} CH_4$$

and

$$N \xrightarrow{H_2} NH \xrightarrow{H_2} NH_2 \xrightarrow{H_2} NH_3.$$

Simple species of these kinds should be detectable in a planetary nebula that has a warm, dense, postshock shell. Water should not be seen if the precursor envelope was carbon-rich. There is insufficient time for reactions between these simple species to occur; all the chemistry is with the major partner, H_2.

Just as the release of molecules from grains in the first two types of region might affect the chemistries in them the desorption of molecules from grains, or the degradation of grains themselves, may contribute substantially to the chemistry in the shocked but not yet ionized parts of the envelope. In IRC + 10216 – a good example of the progenitor of a planetary nebulae – the dust grains are of solid carbon. This is almost certainly amorphous, rather than crystalline diamond or graphite. When carbon grains pass through a sufficiently strong shock they are disrupted into molecules. So the destruction of grains is a source of molecules. In the case of carbon grain destruction, carbon molecules are liberated. Laboratory experiments show that we may expect some linear carbon chain molecules to arise from solid carbon disruption, so this may be the source of the large carbon molecules seen in mature planetary nebulae. Of course, these molecules are destroyed fairly quickly by the intense radiation and by the chemistry in the warm gas. So the large carbon molecules can only appear while the shock is expanding out into the dust-rich envelope.

8.4   Classical novae

Observations show that a star of about the same intrinsic power as the Sun may suddenly brighten optically by an enormous factor, up to about one million, sometimes creating an object capable of detection by the naked eye where none was apparent before; veritably, a *nova stella*. The brightening is accompanied by the ejection of a substantial amount of matter, at high speed.

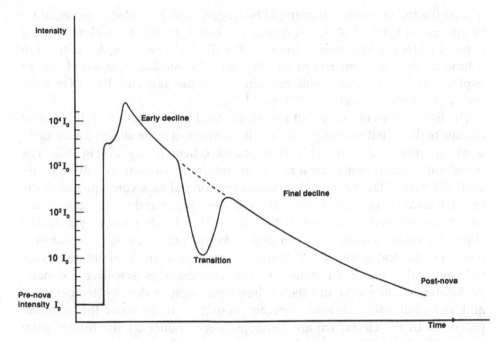

Fig. 8.7. Characteristic light curve of a classical nova. The initial rise takes place within 2 or 3 days, and may take less than 1 day for fast novae. The pause before the final rise may last a few hours for fast novae, up to a few days for slow novae. Novae are at maximum for no more than a few hours (fast novae) to a few days (slow novae). The early decline may take several months. Some novae then pass into a deep minimum lasting for 2 or 3 months, after which they brighten again to follow the late decline. Other novae may undergo large scale oscillations as they pass through the transition region, the oscillations having periods of a week or two. Novae may take a decade or more over their final decline.

A light curve having the characteristic nova shape, which may indicate the formation of dust, is illustrated in Fig. 8.7.

One class of novae, the classical novae, arises in double stars. In a double star system associated with a classical nova, one star has an extended envelope, like those discussed in Section 8.2. The envelope may be so large that the star may be scarcely able to hold on to it by gravity. The other star is a white dwarf. The two stars are in close orbit round each other, the orbital period being typically a few hours. The gravitational force of the white dwarf, so close to the other star, draws matter from the extended envelope of its bloated companion. This matter falls into orbit round the white dwarf and creates a disk of material onto which more gas continually streams from the companion. The gas that feeds the system comes from a cool circumstellar envelope, and has chemical abundances appropriate to that situation.

Frictional effects gradually slow the gas in the disk, and so this relatively hydrogen-rich gas falls slowly toward the white dwarf until the inner ring of the disk grows in thickness and becomes sufficiently heated by the friction to trigger the thermonuclear burning of hydrogen (which requires a temperature of 20 million kelvins) at the surface of the white dwarf. As nuclear burning evolves further, temperatures around 100 million degrees may be attained in a time as short as a quarter of an hour, and the sudden outburst of energy explosively ejects considerable amounts of matter (around 10^{-4} of a solar mass) at speeds up to about $3000 \, \text{km s}^{-1}$.

The light curves of many (but not all) classical novae show that the gradual decline of the visual intensity with time is interrupted by an abrupt transition to a much lower intensity. This is indicated schematically in Fig. 8.7. The transition is accompanied by a rise in the infrared emission, exactly in phase with the optical fading, and is therefore interpreted as a consequence of the rapid formation of dust (which absorbs the optical light and reemits the energy as infrared radiation) in the ejecta. This implies that a substantial chemistry must be occurring prior to dust formation. It now is clear that in at least one classical nova, NQ Vulpeculae, emission in the near infrared from CO molecules has been detected. The chemistry is sensitive to many parameters of the ejecta and the environment, such as density, temperature, and radiation field. Therefore, by attempting to understand the chemical processes in the ejecta, we are developing constraints on the fundamental description of the nova.

The chemistry in the ejecta of classical novae is quite unlike any other chemistry that we have considered so far. Gas densities and temperatures are high so collisions between atoms are very frequent, but radiation fields are very intense so that photodissociation is very fast, and timescales are very short: we are concerned with changes occurring in days or even seconds, rather than thousands to millions of years, as is the case for most other regions and objects described in this book. Nevertheless, the emission features observed in the infrared indicate that chemistry has occurred. For NQ Vulpeculae the infrared emission bands attributed to CO appeared within three weeks of the outburst: this was still three weeks earlier than the transition in the light curve which indicated dust formation. So chemistry *preceded* dust formation, and dust contributed neither to shielding nor to catalysis.

We have to consider, therefore, how chemistry occurs in the ejecta. Although the gas is initially ionized (by the shock wave driven by the explosions and by which the ejecta are accelerated), the densities are high enough that ions and electrons recombine rapidly. We are therefore dealing with a mass of largely neutral gas moving with high velocity away from a very bright star. The intense radiation from the star, unimpeded by dust, 'eats' its way into the receding ejecta ionizing the gas, but at a low speed as measured from the moving ejecta. Therefore, while the ejecta have moved quite far, the radiation has only been able to 'eat' partially through it. However, as the ejecta move out, the density falls and the rate of recombinations of ions and electrons also falls. The radiation then 'eats' out into the ejecta at increasing speed. More of the radiation capable of ionizing carbon atoms can then ionize

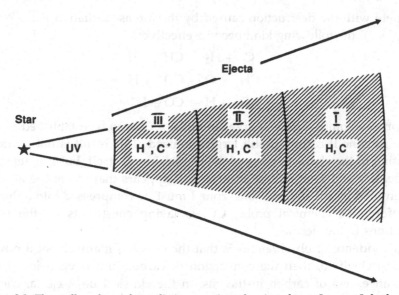

Fig. 8.8. The stellar ultraviolet radiation eats into the ejected gas. In zone I, both hydrogen and carbon are neutral; in zone II only the carbon has been ionized; while in zone III, closest to the star, hydrogen and carbon are both ionized.

oxygen or hydrogen, and as the radiation penetrates it has a characteristic 'bite' shown in Fig. 8.8. Furthest from the star in Fig. 8.8 is zone I, a region which has not yet been ionized. Closest to the star, in zone III, all atoms are ionized, while in between, zone II, only carbon is ionized.

Very soon after the outburst, the radiation has not penetrated very far, so that zone I is close to the star, and densities are high enough for three-body collisions to form H_2

$$H + H + H \rightarrow H_2 + H.$$

Since the ultraviolet radiation field has not yet penetrated, a substantial amount of H_2 builds up. However, collisions can also destroy H_2 in the reverse reaction, by shifting H_2 molecules up the ladder of vibrational states until they reach the top and fall apart.

After a few days the gas is swept by the carbon ionizing front, and is now in zone II. The electrons produced from the ionization of carbon open up an additional H_2 formation process

$$H \xrightarrow{e^-} H^- \xrightarrow{H} H_2$$

which takes over from the three-body reactions as the density falls (see Section 4.1). This helps to maintain a sufficient concentration of H_2 for it to be self-shielding, i.e. the radiation needed to destroy H_2 is entirely used up, so that H_2 in the interior of the region survives. However, because collisional dissociation is so effective at the high temperatures in the ejecta, H_2 is never more than a minor constituent. Nevertheless, its presence is important in allowing further chemistry to proceed. Without H_2, no alternative routes are fast enough to compete with the destruction caused by the intense radiation field. With H_2, schemes of the following kind become effective:

$$C^+ + H_2 \rightarrow CH^+ + H$$

$$CH^+ + O \rightarrow CO^+ + H$$

$$CO^+ + H \rightarrow CO + H^+$$

and proceed quite rapidly. Large amounts of CO can be established in zone I, while in zone II carbon is still largely ionized. This result is not particularly sensitive to the characteristics of the classical nova itself, but it is sensitive to the density and temperature in zone I. It appears that to get the amount of CO implied by the observations, zone I must be compressed into a thin shell. Therefore, the chemical probe, CO, is laying constraints on the physical conditions in the ejecta.

The evidence of observations is that the ejecta of many classical novae are carbon-rich. If so, then the conversion of carbon and oxygen into CO will leave an excess of carbon in the gas. In the classical nova ejecta, chemistry produces large hydrocarbon molecules which may be the precursors of carbonaceous dust.

Selected references

Glassgold, A E, Mamon, G A and Huggins, P J: 'The formation of molecules in protostellar winds', *Astrophysical Journal*, vol. 373, p.254 (1991).

Howe, D A, Millar, T J and Williams, D A: 'Chemistry in a protoplanetary nebula', *Monthly Notices of the Royal Astronomical Society*, vol. 255, p.217 (1992).

Jura, M: 'Chemistry in the circumstellar envelopes around mass-losing red giants', in *Molecular Astrophysics – A Volume Honouring Alexander Dalgarno*, ed. T W Hartquist, Cambridge University Press, Cambridge (1990).

Rawlings, J M C: 'Chemistry in the ejecta of novae', *Monthly Notices of the Royal Astronomical Society*, vol. 232, p.507 (1988).

Rawlings, J M C, Williams, D A and Canto, J: 'Chemistry in T-Tauri winds', *Monthly Notices of the Royal Astronomical Society*, vol. 230, p.695 (1988).

Shu, F H: *The Physical Universe – An Introduction to Astronomy*, University Science Books, Mill Valley, California (1982).

9

Astronomical masers near bright stars

Now my own suspicion is that the Universe is not only queerer than we suppose, but queerer than we can suppose.

J B S Haldane

'Laser' is a familiar acronym for *Light Amplification* by the *Stimulated Emission of Radiation*. The acronym 'maser' is less commonly seen or heard but is closely related; the 'M' in 'maser' stands for *Microwave*. Masers work on the same physical principle as lasers, but amplify microwave, millimetre and radio radiation rather than visible light.

The special characteristic that all masers share is that each is very brilliant over an extremely narrow frequency range and dark at other frequencies. Astronomical masers were discovered even as the Hollywood industry was becoming fascinated with their terrestrial light-amplifying counterparts. These astronomical masers are naturally occurring examples of tremendous radiation amplifiers and exist in the vicinities of the brightest young stars (e.g. they are found near young stars in the Orion region; see Section 6.6) and in the outflows from highly evolved stars which are shedding their outer envelopes (see Section 8.2). They are not only real but considerably more magical than the fictional inventions of the film industry.

We describe the amplification of radiation by stimulated emission in this chapter and consider the chemistries that make possible the existence of some of the different types of astronomical maser. The main question that we ask is this: how do these very special types of source come into existence?

9.1 Stimulated emission

Figure 9.1 shows the different ways in which radiation at the frequency corresponding to the energy separation between two levels of a molecule can

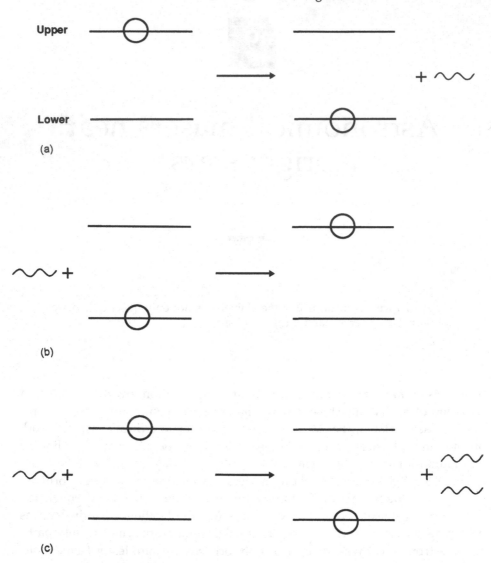

Fig. 9.1. Radiation interacting with two levels of a molecule. (a) The higher energy level is occupied initially but is depopulated, on a timescale that characterizes that level, by the emission of radiation, thus increasing the population of the lower level. (b) The lower level population is decreased by the absorption of a photon giving rise to the population of the upper level. (c) Radiation incident on a molecule in the excited state stimulates its radiative depopulation resulting in the amplification of the radiation field.

interact with that molecule. Three possible types of interactions can take place:

(a) an excited molecule can spontaneously emit a photon, and populate the lower level;

(b) a molecule in the ground level can absorb incident radiation to become excited; and

(c) emission from an excited molecule can be stimulated by incident radiation. If most molecules in a column of gas are in excited levels, then the stimulated emission process will be more rapid than the absorption process, and the intensity of the radiation will be amplified as it passes through the column.

9.2   Populating the upper level – pumping the maser

A maser can only operate when the population in the excited level is high enough relative to that in the lower level. Therefore, the central problem in constructing a maser or a laser is the development of a means of achieving an overpopulation of the upper level. Clearly, amplification of radiation corresponding to transitions between two particular levels leads to a net decrease in the population of the upper level, so the population of the upper level must be replenished by some means. Collisions between molecules that lead to direct population of one level by the depopulation of the other level do not give rise to a sufficient overpopulation of the upper level. At best, even at high temperatures, collisions causing transitions between the two levels lead to relative populations of the two levels that result in no loss, but also in no sustained rate of gain, of power in the emission feature. Hence, masing and lasing cannot occur in a molecule having only two energy levels.

However, all molecules have more than two energy levels. If radiative and collisional transitions to and from a third level are involved in the populating and depopulating of the two levels between which the masing transition occurs, then the necessary large population of the higher level involved in the masing can, under some circumstances, be established. Let us consider a molecule with just three energy levels. (Of course, a real molecule has many more levels, but only three need be involved.) The types of transition between those three levels are shown in Fig. 9.2.

Using the notation of Fig. 9.2, then, our claim for a two-level molecule is that C_{01} (which represents the rate at which collisions excite a molecule from level 0 to level 1) is never large enough compared to C_{10} (which represents the rate at which collisions deexcite a molecule from level 1 to level 0) to enable R_{10} to be greater than R_{01}. This is a consequence of a quantum mechanical principle called *detailed balance* which relates the rates of forward and backward reactions involving two levels. A similar law was discovered before the development of quantum mechanics through the use of statistical mechanical arguments.

However, in a three-level system there are ways in which the population of level 1 can become sufficiently large compared to that of level 0 for masing to occur. We shall consider one simple case. Assume that a strong source of radiation gives rise to transitions between levels 0 and 2, and neglect the effects of all collisions. Level 2 then attains a nonnegligible population which is lost by radiative transitions to the lower levels. In many real molecules, R_{21} is greater than R_{20}; in such systems the population of level 2 gives rise to an enhancement in the population of level 1 with respect to level 0. When the enhancement is large enough, radiation having the frequency corresponding to the energy

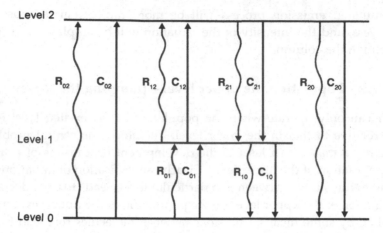

Fig. 9.2. Three levels involved in maser pumping. Wavy lines indicate radiative transitions with rates labelled $R_{i,j}$ where i and j run through 0, 1, and 2 which designate the levels. Straight lines represent transitions caused by collisions involving no photons; collisional rates are labelled $C_{i,j}$. The first digit always gives the initial level and the second always gives the final level.

separating levels 1 and 0 will be amplified by its interaction with the molecules. This type of 'pump', creating a masing relative population between levels 1 and 0, is called a *radiative* pump.

No collisions are involved. However, collisional pumps also exist. For instance, collisions can also populate level 2. However, radiative decays can be more important than collisional deexcitation; then if R_{21} is greater than R_{20}, the enhanced relative population of level 1 to level 0 can result in masing.

The particular pumping mechanism described in the previous paragraph depends on collisions and on the nature of the radiative decay of level 2. However, purely collisional pumps also operate for some systems. For them to work C_{21} must be greater than C_{20}. When a relative level population can lead to masing, a *population inversion* is said to obtain.

9.3   OH masers near young stars

The OH molecule is usually observed at a radio wavelength of 18 cm which corresponds to transitions between sublevels of the lowest rotational level in the lowest vibrational level of OH. (This sort of splitting of a rotational level does not occur for all molecules. However, splitting of the lowest level of OH into sublevels occurs because the electrons in OH have a total angular momentum that can have two different values of its component along the internuclear axis, and because the nuclear spin of the hydrogen can interact in two different ways with electrons. These are details with which we need not be concerned.) The OH 18 cm emission from the brightest young stars (which are as much as 10 thousand times brighter than the Sun) still enshrouded by dense dust and gas is found to be clumped in spots at distances of 10^{14}–10^{15} m from the stars. There are many tens of spots around a single star and each has a size of 10^{12}–10^{13} m (compare this with the distance of the Earth from the Sun, which is about 10^{11} m) and emits at a power in the region of one millionth of the entire luminosity (about 10^{28} joules s^{-1}), summed over all wavelengths, of the Sun. Such tremendous powers in individual radio features, each of which is around 10 thousand million times stronger than a feature emitted under 'normal' conditions at 100 kelvins, can only result from masing.

Theoretical models of the pumping involve collisionally induced excitation and radiative excitation (by the absorption of infrared radiation with a wavelength of about 100 micrometres) to high rotational levels of OH. These high rotational levels subsequently decay preferentially to the upper level involved in a masing transition. The sources of the infrared photons are the hot dust (at a temperature of about 100 kelvins) in the maser regions, and the radiation emitted by the OH molecules themselves. The dust is heated by the radiation of the nearby massive stars. The pumping models are based on detailed information about the properties of OH in collisions with H_2 and in interactions with infrared and radio radiation; an impressive amount of quantum mechanical calculation and clever laboratory investigation was necessary before these very complex astronomical pumping models could be constructed. The results of the theoretical pumping studies permit estimates of the H_2 and OH number densities in the masing regions to be made; they are roughly 10^{13} m^{-3} and 10^7–10^8 m^{-3}, respectively. Therefore, about 1–10 per cent of the oxygen in the masing regions is in the form of OH.

How can so much OH be formed? Why do the spots have the sizes observed? The spots in a maser source appear to be almost (though not totally) at rest relative to their central star. This constrains the answers to these questions; for instance, any shock associated with a slowly moving OH maser cannot by itself have heated the gas sufficiently for the reaction of O with H_2 to form OH to have become rapid enough to be a major source of OH.

In a masing region where the number density is about 10^{13} hydrogen nuclei m^{-3}, then over a length of 10^{13} m dust grains in the gas absorb about enough of the incident radiation from the nearby star that the chemistry begins to switch from one that is dominated by the stellar photons incident on molecular gas to one similar to that of a dark region. The comparability of this length scale and the size of an OH maser spot suggests that the maser spots are in the transition zones between those parts of a cloud just beyond the edge of the sphere that is ionized by the stellar radiation, where the chemistry is strongly affected by the photons, and those parts of a cloud which are essentially dark. Figure 9.3 shows the possible locations of OH masers. We shall refer to the chemistry as being photon dominated in those regions of cloud where the removal of many molecular species is dominated by photon absorption processes rather than reactions with atomic, ionic, or other molecular species. The chemistry of photon dominated regions is of interest in many contexts. The diffuse interstellar clouds (see Section 5.6) are low-density photon dominated regions. There are zones in the Orion

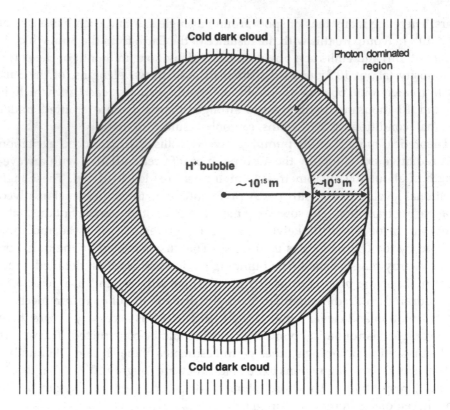

Fig. 9.3. Photon dominated region model of an OH maser region. Stellar radiation that can ionize neutral hydrogen atoms is absorbed near the star. The boundary of the ionized region is sharp. Photons dissociate some but not all of the molecular hydrogen in the shell with a thickness of about 10^{13} m around the boundary of the ionized region. The photons affect the chemistry throughout the 10^{13} m shell and heat a sizeable fraction of it to almost 1000 kelvins. It is in this warm gas that the OH is abundant. The pressure must be almost equal on either side of the H^+ bubble boundary; otherwise, the OH would move faster than observations show.

Kleinmann–Low H_2 infrared emission region in which photons are probably modifying the chemistry in the shocks (see Section 6.6) and providing an additional heat source.

Photoionization of grains is an important heat source near to stars, and photoabsorption by molecular hydrogen is an additional heat source, particularly in gas with number densities above about 10^{11} m^{-3}. At low densities the H_2 simply radiates away most of the energy that it absorbs. However, when the density is high collisions occur so frequently that the energy stored in molecules following photoabsorption is released into the random motion of the molecules rather than being radiated away. So at high densities, H_2 is not a coolant, but a trap for stellar radiation. Since the stellar radiation field in the vicinity of the OH masers is about one million times as intense as the normal interstellar background, it is, therefore, a tremendous source of heat.

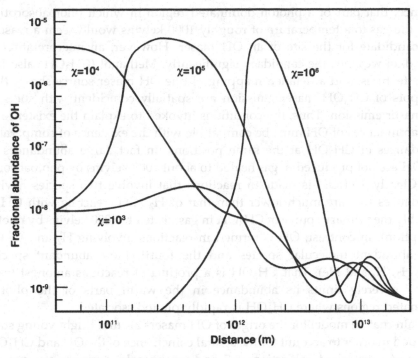

Fig. 9.4. The OH fractional abundance in photon dominated regions with number densities of $10^{13}\,\mathrm{m^{-3}}$. (From T W Hartquist and A Sternberg, *Monthly Notices of the Royal Astronomical Society*, vol. 248, p.48 (1991).)

Heating of the molecular gas in a photon dominated region to temperatures of the order of 1000 kelvins drives the formation of OH through the neutral–neutral reaction

$$O + H_2 \rightarrow OH + H$$

which is slow at low temperatures but rapid at about 1000 kelvins (see Section 5.9). The OH is destroyed by photoabsorption. Figure 9.4 shows the calculated OH fractional abundance in photon dominated regions with number densities of $10^{13}\,\mathrm{m^{-3}}$ and different incident radiation fields; χ is a parameter that indicates how strong the incident radiation field is relative to the normal interstellar background. Notice that for a radiation field about one million times as strong as the interstellar background the OH has a fractional abundance over 10^{-6}, as required by the pumping models, for a length that is comparable to the sizes of OH maser spots. The zone over which the OH is abundant is limited on the side near the star because most of the hydrogen there is atomic; only at some distance into the cloud is there enough H_2 between the point under consideration and the star for the H_2 nearer the star to shield the H_2 at that point from the dissociation inducing ultraviolet radiation. The zone where the OH is abundant is limited on the other side because the dust absorbs most of the photons that must keep the gas hot, and as the heating declines the $O + H_2$ reaction becomes slow.

Hence, that part of a photon dominated region in which photoabsorption heats the gas to a temperature of roughly 1000 kelvins would seem a reasonable candidate for the site of an OH maser. However, an interpretation of other data weakens the candidacy significantly. Methanol (CH_3OH) also has maser features, and a detailed mapping of one OH maser source shows that the spots of CH_3OH maser emission are spatially coincident with spots of OH maser emission. Thus, the conditions invoked to explain the existence of high abundances of OH must be compatible with the existence of comparable abundances of CH_3OH at the same positions. In fact, large abundances of CH_3OH are not produced in gas heated to about 1000 kelvins by photoabsorption. Clearly, CH_3OH is made in reactions that involve two species having abundances that are much lower than that of H_2 (CH_4 reacting with OH is probably the primary source of CH_3OH in gas heated to 1000 kelvins by photoabsorption). In contrast, OH is formed in reactions involving H_2 and O, the most abundant molecular species and the fourth most abundant species (after H_2, H, and He). That CH_3OH is a product of reactions amongst trace species severely limits its abundance in the warm parts of the photon dominated regions where CH_3OH is rapidly photodissociated.

An alternative model for the origin of OH masers around bright young stars is needed in order to account for the spatial coincidence of the OH and CH_3OH masers. As described in Sections 5.7 and 6.5, in cold dark regions species containing elements heavier than hydrogen and helium freeze onto dust grains on which some chemical processing takes place. Specifically, CH_3OH ice is known to form on dust grains and is observed (through the detection of its infrared spectral features in absorption towards bright young stars enshrouded by large amounts of dust) to exist on interstellar grains. H_2O ice is also abundant on grains in dark cold regions. If grains are heated to about 100 kelvins or higher, CH_3OH and H_2O are evaporated and enter the gas. The origin of gaseous CH_3OH in the masers almost certainly must be the melting of CH_3OH ice in a region that has been recently heated.

Heating to release the CH_3OH into the gas of the masers requires that the gas and dust move slowly towards the star, at about several kilometres per second. This speed does not conflict with the observed limits of the motions of OH maser spots relative to the star with which they are associated. Melting of CH_3OH and H_2O ices may result from heating dust grains by the absorption of infrared photons emitted by warmer dust nearer to the star. Alternatively, collisions between large and small grains in the flowing gas may chip off small pieces of ice which subsequently melt through collisions with the gas. The gaseous CH_3OH and H_2O molecules then move slowly towards the photon dominated zone. In the outermost parts of the photon dominated zone (where the temperature is far below 1000 kelvins) the H_2O is photodissociated to form OH; some of the CH_3OH is photodissociated as well. As the material moves through the photon dominated zone the OH and more of the CH_3OH are photodissociated and the abundances of the species drop. By chance, the photodissociation rates of H_2O, OH, and CH_3OH are nearly the same which means that H_2O, OH, and

CH$_3$OH will all be abundant in the same outer part of the photon dominated region. The molecules OH and CH$_3$OH show maser activity there, though H$_2$O does not, because the physical conditions necessary to pump OH and CH$_3$OH masers are similar, but differ substantially from those required to pump H$_2$O masers.

9.4   H$_2$O masers near young stars

Unlike OH masers, H$_2$O masers in the vicinities of young stars show high speeds of up to about 100 km s^{-1}. The masing transition is between two rotationally excited levels and the wavelength of the emitted radiation is 1.3 cm. The pumping mechanism is thought to be collisional, and requires a temperature of several hundred kelvins to operate since the upper level is so excited. The number densities consistent with the pumping mechanism lie between about 10^{14} m^{-3} and 10^{15} m^{-3}. The individual Galactic H$_2$O maser spots have dimensions of about 10^{11} m and powers between 10^{-5} and 10^{-1} of the total solar luminosity. While the Sun's power is spread over a wavelength band comparable to the wavelength of emission, the maser power is confined to a wavelength band that is only about 10^{-5} the wavelength of the emission.

Figure 9.5 shows the distributions of OH and H$_2$O masers in the vicinity of the Kleinmann–Low Nebula (see Section 6.6). The H$_2$O masers belong to two separate populations distinguished by their speeds. A family of low-velocity H$_2$O masers is moving at about 18 km s^{-1} away from the infrared source called IRc2 (which was also referred to as IRS2 in Section 6.6). A smaller number of high-velocity H$_2$O maser spots move at velocities of the order of 100 km s^{-1} away from IRc2. A stellar outflow of 10^{-4}–10^{-2} solar masses per year is thought to drive both types of maser motion; asymmetry in the ambient medium distribution, due perhaps to the existence of a thick disk (see Section 7.1) around the young massive star associated with IRc2, may retard the outflow in some directions, resulting in the low-velocity masers, while permitting it to remain fast in other directions, so giving rise to the high-velocity masers. Shocks near the interfaces between the wind and the ambient medium (see the wind-blown bubble structure described in Section 4.3) probably create the elevated temperatures required for H$_2$O maser pumping.

The fast H$_2$O masers are of special interest, since shocks with speeds as high as 100 km s^{-1} almost certainly cause the dissociation of molecules. A shock with a speed greater than 26 km s^{-1} entering a nonmagnetic medium with a number density in excess of about 10^{11} m^{-3} will dissociate all H$_2$ molecules present; when the shock speed is in excess of about 100 km s^{-1} in a gas without a magnetic field the postshock temperatures are so high that the gas is ionized. Magnetic fields might provide some cushioning and change these critical speeds somewhat, but it is unlikely that the magnetic field is strong enough in the Orion Kleinmann–Low Nebula to prevent a 100 km s^{-1} shock from inducing complete dissociation of H$_2$. Hence, the gas in shocks in which the high-velocity H$_2$O masers form can contain no molecules shortly after the initial heating.

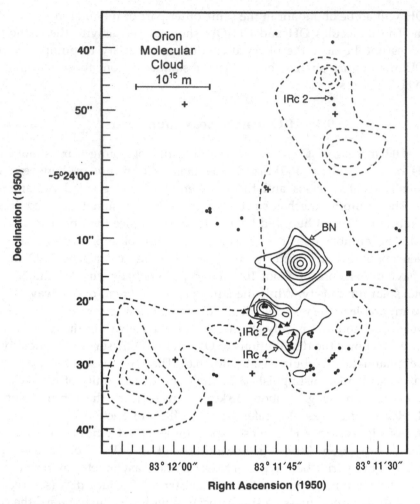

Fig. 9.5. Distributions of masers in the Orion Kleinmann–Low Region. The dashed contours are for H_2 infrared emission while the solid contours are for infrared emission from warm dust. Dashed contours peak on shocks and solid contours peak on the sites of dusty clumps heated by the radiation of a young star. The positions of high-velocity H_2O masers, low-velocity H_2O masers, and OH masers are indicated with crosses, dots, and triangles, respectively. Strong CH_3OH masers are found at the locations marked with squares. (From M J Reid and J M Moran, *Annual Reviews of Astronomy and Astrophysics*, vol. 19, p.231 (1981).)

However, if the ambient medium number densities are around 10^{13} m^{-3}, then high postshock number densities of 10^{14}–10^{15} m^{-3} are produced. The timescale for hydrogen to strike grains and reform H_2 (see Section 5.4) in this very dense gas after it has cooled somewhat, is only about one year. At a flow speed of 100 km s^{-1}, a reformation time of 10^7 s (~year) corresponds to a length of 10^{12} m, rather larger than but roughly comparable to the observed sizes of the fast H_2O maser spots. The approximate equality of the H_2 reformation length-scale and the H_2O maser sizes suggests that the masers are produced in the H_2 reformation region. Figure 9.6 shows schematically how the temperature

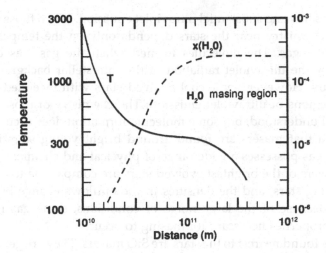

Fig. 9.6. The temperature and H_2O fractional abundance behind a dissociating shock. The temperature (in kelvins) immediately behind a $100\,km\,s^{-1}$ shock propagating into gas with a number density of $10^{13}\,m^{-3}$ is around 100 000 kelvins above absolute zero. All H_2 is rapidly dissociated. As the temperature falls to several thousand kelvins, H_2 is reformed when atomic hydrogen collides with grains. H_2O is formed in the cooler gas. Its presence leads not only to masing but also cooling which balances the heating. The H_2O masers exist where the H_2 reformation occurs. (From M Elitzur, D J Hollenbach and C F McKee, *Astrophysical Journal*, vol. 346, p.983 (1989).)

varies behind a shock in which H_2 reformation occurs. The reformation of H_2 molecules releases energy (since bound H_2 is a lower energy system than two free H atoms) at a rate that maintains the temperature at about 400 kelvins, which lies in the range necessary for the H_2O maser collisional pump to be effective. The H_2O molecules form due to reactions of O and OH with H_2 which are rapid at elevated temperatures (see Section 5.8). H_2O emits in the infrared following the excitation of its rotational levels in the collisional pumping process. This infrared emission provides the dominant cooling that prevents the H_2 formation from maintaining the gas at even higher temperatures.

Unfortunately, the postshock H_2 reformation region model applies only to the fast H_2O masers. The production of conditions favourable for H_2O masing in the slow spots has not yet been understood. The type of collisional pump operating in the slower spots may differ from that which seems to apply to the faster ones.

9.5   Masers in stellar outflows

In Section 8.2, we described the chemistry that occurs in envelopes around very evolved stars of a solar mass, or larger. These envelopes are expanding with velocities of around $20\,km\,s^{-1}$. The outer envelopes of many stars contain more oxygen than carbon; these oxygen-rich envelopes contain masers.

The formation of molecular species in these outflows is relatively straightforward if the flow timescales are long compared to the chemical timescales and if the stars have no substantial ultraviolet emission. When those conditions are met the chemistry near the stars depends only on the temperature and pressure of the gas. The molecules formed when the gas was denser are dissociated by the ultraviolet radiation of the interstellar background as the outflow occurs. However, some cool evolved stars with extended envelopes do possess regions of ultraviolet emission. The chemistry of those envelopes is not so well understood, but some molecule formation does occur.

One reason that masers are found around bright young stars is that the surrounding gas possesses a wide range of physical and chemical properties. The total powers of the brightest evolved stars are comparable to those of the brightest young stars, and the densities in the outflows change by orders of magnitude. Because of the large range of conditions, some gas is likely to possess the properties necessary for masing to occur.

The masers found nearest to the stars are SiO masers. They are generally seen at radii of about 10^{11} m, comparable to the distance between the Earth and the Sun. They probably contain about 10^{-6}–10^{-5} times the mass of the Sun. The pumping models for SiO masers are not able to account for the great power emitted in the observed masing transitions between the lowest and first excited vibrational levels of SiO, but they indicate that the number densities in the maser spots are as high as 10^{16}–$10^{18} m^{-3}$. Thus, the SiO maser spots have (possibly greatly) enhanced densities relative to the surrounding gas in the outflow, which has a density that is several times, or even a hundred times, lower. The clumps in which SiO masers form may survive for long times, though at the lower number densities of $10^{11} m^{-3}$, during various phases of stellar outflow, and well into the planetary nebulae stage (see Section 8.3). In some places in the envelopes SiO formation leads to silicate dust formation.

Stellar envelope H_2O masers are seen at distances of roughly 10^{12} m from the stellar centres. Their physical properties are probably similar to those of the H_2O masers around bright young stars. As the outward flow continues, H_2O is photodissociated by the interstellar radiation field and OH is formed. Stellar envelope OH masers exist at distances of 10^{14} m from the central stars.

Hence, maser emission marks the passing of some stars from activity to long slow deaths as white dwarfs. Maser emission may also signal the birth of some of the most brilliant newcomers to the stellar population.

Selected references

Elitzur, M: *Astronomical Masers*, Kluwer Academic Publishers, Dordrecht (1992).

Elitzur, M, Hollenbach, D J and McKee, C F: 'H$_2$O masers in star forming regions', *Astrophysical Journal*, vol. 346, p.983 (1989).

Hartquist, T W, Menten, K M, Lepp, S and Dalgarno, A: 'On the spatial coincidence of hydroxyl and methanol masers', *Monthly Notices of the Royal Astronomical Society*, vol. 272, p.184 (1995).

Moran, J M: 'Masers in the envelopes of young and old stars', in *Molecular Astrophysics – A Volume Honouring Alexander Dalgarno*, ed. T W Hartquist, Cambridge University Press, Cambridge (1990).

10

Supernovae: fairly big bangs

―――

Not till the fire is dying in the grate
Look we for any kinship with the stars.

George Meredith

The Big Bang that created the Universe filled it with hydrogen. The simple chemistry that developed influenced the formation of the galaxies (see Chapter 4). Stars more massive than about eight solar masses in those galaxies rapidly consume their primary fuel, hydrogen, and undergo enormous explosions called *supernovae*, seeding the galaxies with the products of thermonuclear burning which produces mainly carbon, oxygen, nitrogen, silicon, and iron. The injection of heavy metals such as silicon and iron into interstellar space is almost entirely due to supernovae, whereas mass loss from low-mass stars (see Chapter 8) is also important for other elements, especially carbon, oxygen, and nitrogen. Since iron is such an important component of our blood, the funeral fires of the more massive stars have made our lives possible. In this chapter we discuss the chemistry occurring in the ejecta of a supernova, and how we can learn about the physical state of the ejecta from the chemistry.

10.1 Introduction to supernovae

Through supernovae, galaxies become more enriched in heavy elements as they grow older, so stars formed recently have higher abundances of these elements in them than very old stars. All galaxies have many low-mass stars which evolve only slowly and so are very long lived, many of them being nearly as old as the galaxy of which they are members. But more massive stars have shorter lives; though they have more fuel to burn, they burn it many times faster than a low-mass star. A really bright star may have 50 times as much fuel as the Sun, but burns it at a rate more than a million times faster so it can only live about 10^{-4} times the life of the Sun. All the bright stars we can see are therefore very recent additions to our Galaxy. They are made of interstellar gas that is steadily becoming richer in oxygen, carbon, nitrogen, and other heavy elements. At the end of their lives, these massive stars put back into the interstellar medium nearly all their mass in supernova explosions, so these stars are made of matter which continually cycles into and out of stars, mixing with the interstellar gas and enriching it with heavy elements. This cycle is driven by star formation; the enrichment of the interstellar medium and the mixing of old and new interstellar gas are driven by supernova explosions.

Theoretical understanding of supernovae became well enough developed during the 1980s to explain how a substantial fraction of a star's mass could be ejected at speeds of several thousand kilometres per second. However, a comparison of theory with observation was impossible because no supernova had occurred in a suitable time and place. The last one observed in our own Galaxy was in 1604, and there were then no instruments capable of studying it. Supernovae occur in other galaxies but these are usually very distant so that we cannot discern the detail.

However, on 24 February 1987, one of the most spectacular astronomical events was seen. On that day a supernova, SN 1987A, was detected in the companion galaxy to the Milky Way, the Large Magellanic Cloud. In fact, a burst of neutrinos had been registered the previous day in two neutrino detectors designed primarily to count neutrinos produced in the Sun. These neutrinos originated in SN 1987A and had been travelling at the speed of light for 170 000 years. Figure 10.1(a) shows the Large Magellanic Cloud before and after the explosion, and Fig. 10.1(b) identifies the actual star that exploded. The light received from SN 1987A reached its greatest optical brightness on 20 May 1987, when its intrinsic power output was as high as several hundred million stars like the Sun. The light curve of SN 1987A is shown in Fig. 10.2. Here, at last, was an opportunity to study the processes at the end of the life of a massive star, and the injection of matter into interstellar space. The amazing result is that, though there have been some surprises, the overall theoretical picture of a supernova explosion has been confirmed by the observations.

It might seem unlikely that chemistry would play a role in the explosion of such enormous magnitude. The total energy is 10^{46} joules; by comparison in one year the Sun radiates a total of 10^{34} joules, a millionth of a millionth of the supernova. This is really big stuff! Yet chemistry seems ubiquitous. Where there is an opportunity, molecules will form. In fact, several varieties of molecule have been detected in SN 1987A, viz. CO and SiO. The chemistry that occurs in the ejecta of supernovae must be quite different from that in interstellar clouds. It has its own intrinsic interest. But that it occurs at all, and that it produces observable amounts of molecules, gives insight into the effects of the explosion and conditions in the ejecta. The chemistry is a useful probe, though it does not of itself modify the physical conditions in supernovae ejecta.

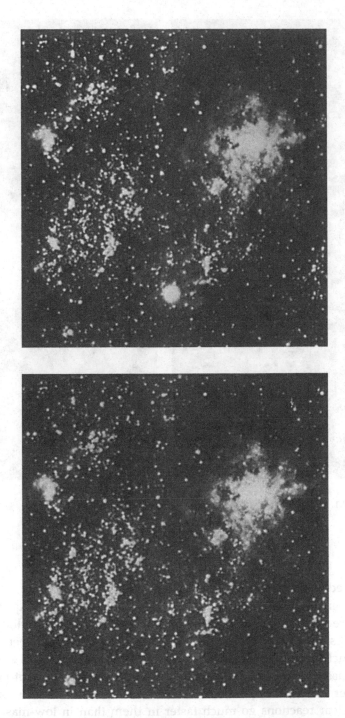

(a)

Fig. 10.1. Before and after: the appearance of SN 1987A: (a) shows the entire Large Magellanic Cloud, (b) indicates the star that became the supernova.

(b)

Fig. 10.1. (*cont.*)

10.2   What happens in a supernova

As described at the beginning of Chapter 8, the basic process powering stars
is the conversion of hydrogen nuclei to helium nuclei. This requires a high
temperature (the temperature at the Sun's centre is over 10 million kelvins)
so that collisions can overcome the electrostatic repulsion between positively
charged nuclei, but when this nuclear fusion does occur there is a big energy
release. Massive stars, say those with masses greater than about eight Suns,
have higher internal temperatures than those of low-mass stars, so that the
thermonuclear reactions go much faster in them than in low-mass stars. As
in lower-mass stars, the reactions go fastest at the centre of a massive star,
so that is where hydrogen is first used up and the abundance of helium
increases. When all the hydrogen in the core is exhausted, the heat source
is quenched and – without this source of energy – the core is compressed

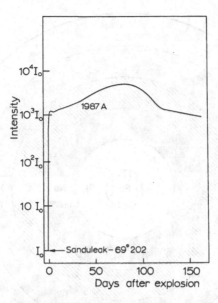

Fig. 10.2. Optical light curve of SN 1987A. Sanduleak −69° 202 is the name of the star that exploded in a supernova.

under the weight of the star. The temperature rises in this compression, and becomes high enough to ignite a new sequence of fusion reactions in which helium nuclei form carbon. We considered at the beginning of Chapter 8 how a star with a mass less than about six solar masses develops a core primarily of carbon surrounded primarily by helium and hydrogen shells. During this phase of evolution a low mass star loses much of its mass in an outflow of speed of roughly 10–20 km s^{-1}, and a white dwarf is eventually left behind. A star more massive than six solar masses also develops a carbon core, but the mass of that core exceeds the maximum possible mass of a white dwarf. Gravity compresses the carbon core to an extent much greater than possible in a low-mass star. This gravitational compression leads to high enough temperatures to overcome the electrostatic barriers between nuclei larger than carbon and helium, and a star more massive than six solar masses goes through several more phases of evolution forming progressively larger nuclei until iron, the most stable of all nuclei, becomes abundant. Figure 10.3 shows the 'onion-like' structure of shells surrounding the iron core at the end of the nonexplosive nuclear burning phase of a massive star's life.

In the core, iron nuclei, which have the greatest binding energy per nucleon of all elements, do not fuse together to release energy; it takes an input of energy to get them to combine. Suddenly, the star has no nuclear energy source at the centre, and must cool unless some other energy source is available. That alternative source is gravitational collapse which generates heat while the core collapses to form a neutron star. The collapse of the core of a star with a mass exceeding about eight solar masses gives rise to such a rapid generation of heat that a supernova occurs. The layers of the star outside

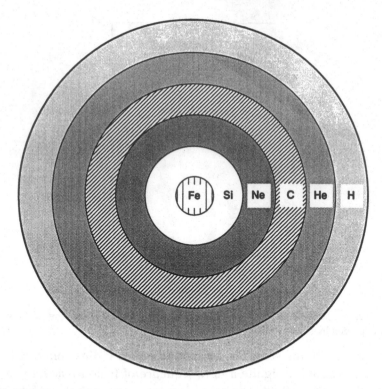

Fig. 10.3. The structure of the star before it became a supernova.

the iron core also fall inwards, only to impact on the solid neutron star and bounce off it. In the transient heating accompanying the bounce further nuclear reactions occur, and new elements are created, some of which are heavier than iron, including some that are radioactive. It is the bounce which causes the ejection of the outer layers of the star.

This general picture of a supernova has been confirmed by observations of SN 1987A. But there are many questions to which we would like the answers. Do the 'onion layers' lift off intact in a stratified fashion, or is there mixing between the layers? What are the physical conditions in the ejecta? What heats the ejecta and how do they cool? To what extent does the underlying neutron star affect the ejecta? Will dust form in the ejecta? If so, will the dust obscure the underlying neutron star?

The discovery in the spectrum of SN 1987A of two infrared lines of vibrational emission from CO, at a time of 112 days after detection of the outburst, was the first identification of molecules in supernova ejecta. Clearly, some kind of chemistry was occurring, even in this hostile environment. The molecule SiO appeared also to be present, as did H_3^+. The latter molecule plays a key role in dark cloud chemistry (see Section 5.5). Does it have an important part in supernova chemistry?

We describe in the next section the kind of chemistry that may occur and show how the presence of these few types of molecule helps to limit the range of possible physical conditions in the ejecta.

Table 10.1. *The inner and outer ejecta of a supernova.*

	Inner	Outer	
Velocity of expansion ($km\,s^{-1}$)	700	2000	
Atomic number density (m^{-3})	10^{15}	10^{15}	
Temperature (kelvins)	2000	6000	
Percentage abundances			(Interstellar
(compared to the interstellar medium)			medium)
hydrogen	0	45	(90)
helium	54	53	(9)
oxygen	33	1	(0.1)
carbon	5	0.3	(0.03)

10.3 Supernova chemistry: a hydrogen-poor environment

In Chapter 4 we saw that the chemistry in the Early Universe was almost entirely that of hydrogen. In the interstellar medium and star forming regions (Chapters 5 and 6) there are traces of heavy elements (especially O, C, S, and Si) and a chemistry of great variety and importance arises, in which reactions with H_2 are the most frequent type of reaction. However, in supernovae ejecta hydrogen is no longer the dominant element: although in the outer part of the ejecta some hydrogen remains so that its abundance is comparable to that of helium, in the inner part of the ejecta almost all of it has been processed into heavier elements. Abundances and physical conditions in the inner and outer parts can be calculated from a theoretical model, the details of which are given in Table 10.1. There may be some mixing, caused by the explosion, between the inner and outer parts. Both regions of the ejecta are hydrogen-poor compared to any other astrophysical situation, so the chemistry will be unusual. Dust grains, as in the ejecta from novae (see Section 8.5), will form only after the chemistry becomes effective. Hence, we shall consider chemistry in dust-free ejecta of supernovae.

Let us look at the chemistry that might lead to CO in the inner ejecta, where hydrogen plays no role. CO was detected in SN 1987A about three months after its outburst, by which time the density had declined to about $10^{16}\,m^{-3}$ so that three-body reactions did not operate. The temperature in the ejecta at this time was several thousand kelvins.

Without hydrogen, the chemistry in such ejecta depends on rather slow associations such as

$$C + O \rightarrow CO + radiation$$

and

$$C + e^- \rightarrow C^- + radiation$$

followed by

$$C^- + O \rightarrow CO + e^-.$$

149

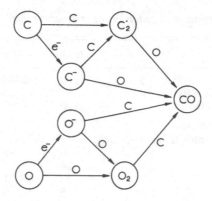

Fig. 10.4. Chemistry of CO formation in a supernova.

A summary of the likely routes is shown in Fig. 10.4. Such a chemistry would be very slow in interstellar clouds. In supernova ejecta the density is about one billion times larger, so that even though reactions of C with O and with e^- proceed in less than one in a million collisions, the number of reactions is still very large. Even in the simple chemistry involving only carbon, oxygen, helium, and electrons there are many routes to CO.

Of course, these formation routes compete with various routes that destroy CO, and in supernova ejecta conditions are probably quite severe. Let us consider what the major loss mechanisms are likely to be. After the initial rise and fall in the light curve (as shown in Fig. 10.2) SN 1987A showed a steady decline which matched exactly in its rate the radioactive decay of ^{56}Co, a radioactive isotope of cobalt created in the transient heating at the 'bounce'. These ^{56}Co nuclei emit powerful gamma rays which are the main energy input to the supernova ejecta. First, the gamma rays collide with electrons in the gas, sharing their energy with them, so that the electrons move very quickly. These fast electrons collide with the abundant helium atoms, ionizing some of them, and raising others to excited states. When these excited helium atoms radiate, energetic radiation in the far ultraviolet region is generated. Thus, the gamma ray energy source generates fast electrons, causes ionization, and stimulates a diffuse ultraviolet radiation field throughout the ejecta. So a variety of destruction processes oppose the formation reaction network in supernova ejecta. Fast electrons can tear molecules apart

$$CO + e^-(\text{fast}) \rightarrow C + O + e^-.$$

Negative and positive ions can mutually neutralize each other, e.g.

$$C^- + O^+ \rightarrow C + O.$$

Negative ions are rather easily destroyed by near infrared radiation

$$C^- + \text{radiation} \rightarrow C + e^-.$$

As in the interstellar medium, CO is destroyed by He^+

$$CO + He^+ \rightarrow C^+ + O + He$$

but here He^+ arises indirectly from the ^{56}Co gamma rays, not from the cosmic rays. The most severe attack on CO may be photodissociation by the ultraviolet radiation field

$$CO + radiation \rightarrow C + O$$

for there is no hiding place for the CO, unlike in interstellar clouds; there are no dust grains to shield the gas. The radiation is intrinsic, internal to every parcel of gas since ^{56}Co – the source of the radiation – is mixed throughout the gas. So CO cannot shield itself here, as it does in molecular clouds.

What can we learn from the observations of CO in SN 1987A? The major uncertainty in the chemistry seems to be the strength of the ultraviolet radiation field in the ejecta. The fact that CO molecules have been observed places a constraint on the intensity of the CO dissociating radiation. Somehow, in the cascade of energy from the gamma rays, not too much energy could have been converted into radiation capable of destroying CO. Hence, the detection of CO has created an as yet unsolved problem in understanding the energy deposition – heating process in the ejecta.

The detected CO seems to have been located in the inner part of the ejecta where H was absent. The expansion velocity in this region, though large, was not as great as that in the outer region which has – of course – moved further. Table 10.1 gives theoretical model parameters typical of the inner and outer ejecta. The model results indicate that the outer part had substantial amounts of hydrogen (see Table 10.1), though by interstellar standards was still hydrogen-poor. The outer part of the ejecta showed other lines, including some emitted by atomic hydrogen when electrons and protons recombine. Hence, observations confirm the expectations of the models concerning the composition of the outer part of the ejecta. This hydrogen containing region allowed the formation of H_3^+ in sufficient quantities that it has been observed. Two infrared lines are attributed to H_3^+. The lines were detected about four months after outburst, and reached peak intensity some four months later. They appear to have been lines of the vibrational spectrum of H_3^+ emitted in a gas expanding at more than $3000\,km\,s^{-1}$. This is consistent with information from the hydrogen lines.

The chemistry that may form H_3^+ in a supernova seems to be similar to that in the Early Universe. It is illustrated in Fig. 10.5. An additional route to H_2^+, not possible in the Early Universe, is also available in supernova ejecta. It involves excited H atoms, H^*,

$$H + H^* \rightarrow H_2^+ + e^-.$$

In no other environment that we have considered does H^* play an important chemical role; this is because the radiative lifetime of H^* is so short that in lower density environments it emits a photon on a timescale much shorter than the timescale for hydrogen in its ground state to be excited in a collision to produce H^*.

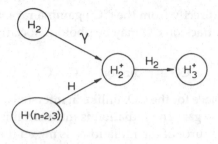

Fig. 10.5. H_3^+ formation in supernova ejecta.

What can we learn from this observation of H_3^+ in SN 1987A? The amount of H_3^+ that can be made is not much smaller than that implied by the observation. It seems possible that if the gas is clumpy rather than uniform, the amounts of H_3^+ in the clumps would increase enough so that the total emission would match that observed. It appears that the chemistry is telling us something about conditions in the outflow. Most gas we observe in astronomy is clumpy, on all scales. Supernova outflows may also be clumpy.

Selected references

Lepp, S, Dalgarno, A and McCray, R: 'Molecules in the ejecta of SN 1987A', *Astrophysical Journal*, vol. 358, p.262 (1990).

Liu, W, Dalgarno, A and Lepp, S: 'Carbon monoxide in SN 1987A', *Astrophysical Journal*, vol. 396, p.679 (1992).

Rawlings, J M C and Williams, D A: 'Chemistry in supernova 1987A', *Monthly Notices of the Royal Astronomical Society*, vol. 246, p.208 (1990).

Shu, F H: *The Physical Universe – An Introduction to Astronomy*, University Science Books, Mill Valley, California (1982).

11

Molecules in active galaxies

Some concentration of the luminosity of the Milky Way Galaxy is found towards the Galactic Centre, but if the material within a central sphere of radius about several thousand light years were removed, the overall power of the Galaxy would not be diminished significantly. (Just for comparison, the Sun's orbit around the Galactic Centre has a radius of about 25–30 thousand light years.) However, a small percentage of galaxies would be significantly dimmed if the material in their central several thousand light years were removed. These galaxies are referred to as active galaxies. In many cases the distributions of molecular emissions in active galaxies differ substantially from that in the Milky Way, so we can hope to learn something about those galaxies by considering the molecules they contain.

Many different types of active galaxy exist. In some cases, a burst of massive star formation in the central volume produces enough radiative power to compete with or even dominate the brightness of the stars distributed through-out the rest of the galaxy; such a galaxy is called a starburst galaxy. Often no starburst is occurring, and the central engine dominating the galaxy's power is very small in spatial extent; rapid variations in the emission properties of many active galaxies show that the engines are less than light months in size. These compact, though powerful, central engines are believed by many astronomers to consist of matter accreting on to black holes with masses of between one million to one thousand million times that of the Sun.

It is possible that an evolutionary sequence relates some of the different types of active galaxy to one another. If there is an evolutionary sequence, can observations of molecular emission from active galaxies be interpreted to help us to understand how one type of active galaxy develops to become another type? Can the masers in some active galaxies that are many times more powerful than the masers in the Milky Way be studied to learn more precisely how far away those galaxies are and to study the expansion of the Universe? What do molecular features in the spectra of quasars, the most distant and intrinsically most powerful active galaxies, tell us about the era of galaxy formation?

11.1 The black hole – accretion disk model of the central engines

Black holes, despite their picturesque name, are no more mysterious than any other natural phenomena; the ways in which matter and energy interact according to a very small number of physical laws to give the amazing complexity and beauty of a star, a planet, or a mind are just as magical. The most easily understood model of the black hole was advanced in the eighteenth century by the Rev. John Michell (a friend of Henry Cavendish), and discussed by Joseph Priestley and subsequently by the Marquis de Laplace. Michell reasoned that just as the Earth has a minimum escape speed above which a space probe going to Mars, say, has to be launched, other astronomical bodies have escape speeds, including some that exceed the speed of light. If the gravity of a body is strong enough that light cannot escape from its surface, then we cannot see that surface, and it appears black.

For a black hole of, say, 10^8 solar masses to form, the matter must reach a density of at least several thousand kilograms per cubic metre (which is less than the average density of typical rocks and about treble that of water) and be contained within a sphere of radius of a few hundred million kilometres, about double the Earth–Sun distance. The formation of such black holes seems to be the natural consequence of stellar collisions that occur in the dense stellar clusters at the centres of the largest, most centrally concentrated galaxies. Once a reasonably massive seed black hole forms, its tidal forces can rip away the extended envelopes (see Section 8.2) of evolved stars in the central cluster of the host galaxy and grow by accreting the stripped-off gas, which forms an accretion disk around the black hole. Even the Galaxy, which is not active according to the definition that we are using, appears to have a central black hole with a mass of about a million times that of the Sun.

Most members of some classes (e.g. quasars) of active galaxy are at distances that are a substantial fraction of the size of the visible Universe and that are measured in billions of light years. Some types of activity in galaxies are seen only in very distant objects and, therefore, appear to belong primarily to the distant past, implying that galaxies evolve as they age. Activity may cease when the fuel supply in the central star cluster of a large galaxy (i.e. many of the cluster's stars) has become too depleted to power the monster at such a furious rate.

11.2 Starburst galaxies

Much of the infrared luminosity of the Galaxy is emitted by dust grains heated by the absorption of starlight. A survey of the sky at infrared wavelengths (about 10–100 micrometres for this mission) conducted in the early 1980s by the Infra Red Astronomical Satellite, IRAS, revealed the existence of galaxies that were very bright at infrared wavelengths, with total luminosities about one million million (10^{12}) times that of the Sun and about ten times that of the Galaxy. The infrared and detected CO emissions were found to be concentrated towards the centres of these galaxies. Spectacular bursts of star formation occurring over periods of about several million years are thought to power the infrared emission of many of these galaxies.

In addition to the continuum infrared radiation from warm dust grains, the starburst galaxies also emit discrete lines at wavelengths of about 2 micrometres, arising from transitions between vibrational levels of molecular hydrogen, H_2. As described in Section 6.6, regions of massive star formation in the Milky Way are also strong sources of such radiation. In the nearby Kleinmann–Low Nebula behind in the Orion Nebula, where star formation is occurring, most of the H_2 emission occurs in shocks driven by the wind of one of the recently formed stars. The ratios of power in the infrared continuum to that in the H_2 infrared lines are somewhat less in a number of the starburst galaxies than in the Orion cloud, possibly indicating that the H_2 emission from some of the starburst galaxies arises in a somewhat different fashion than it does in Orion. For instance, the shocks may be driven by supernova explosions rather than the winds of young stars; young stars and supernovae are likely to exist simultaneously in the starburst galaxies since the largest stars evolve on timescales that are comparable to those over which the star formation is likely to persist. Alternatively, ultraviolet radiation emitted by the stars or X-rays emitted by supernova and their interactions with the interstellar gas surrounding them (or by accretion onto a central 10^8 solar mass black hole) may heat molecular gas so that it produces H_2 line radiation.

The origin of many of the starbursts is thought to be driven by collisions between galaxies. Distortion of a galaxy in such a collision often will give rise to a bar-like distribution of matter in the galaxy. Figure 11.1 shows photographs of some colliding galaxies. Theoretical studies show that the interaction of a bar with molecular clouds that are losing energy in collisions with one another results in the rapid spiralling (i.e. in several rotation periods of the galaxy) of the clouds to the centre of the galaxy. When they reach the centre, their frequent collisions seem to trigger bursts of massive star formation. Further studies of the emissions from many molecular species may result in the eventual construction of a more detailed description of the origin of starburst galaxies than the sketchy one existing at this time. The nature of the interaction of molecular clouds certainly determines the star formation rate, but we know little as yet about these interactions.

11.3   Seyfert galaxies

Many, if not all, of the starburst galaxies have even more concentrated emission sources like those associated with a class of active galaxies called Seyfert galaxies. These Seyfert galaxies possess bright centres with optical features emitted by gas having a wide velocity distribution. (We encountered the concept of motions affecting spectral feature profiles in Section 6.5.) The features from Seyfert galaxies are much wider than those of a couple of hundred kilometres per second, typical of the orbits of clouds in the Galaxy. These features are formed when diffuse gas is heated by the absorption of radiation from a central source (probably an accreting black hole). Any one Seyfert centre possesses a wide range of different types of detected atoms and ions, from neutral hydrogen to neon carrying a charge of $+4$ elementary charges.

Observations of molecular features formed in Seyfert galaxies have been conducted in attempts to understand how starburst galaxies and Seyfert galaxies (or at least a subclass of them) are related. Seyfert galaxies are generally divided into two types. In a type 1 Seyfert some of the optical features from the central regions are broadened by higher-velocity motions while other optical features are relatively narrow. In a type 2 Seyfert all of the centrally emitted optical features are broadened by the same amount.

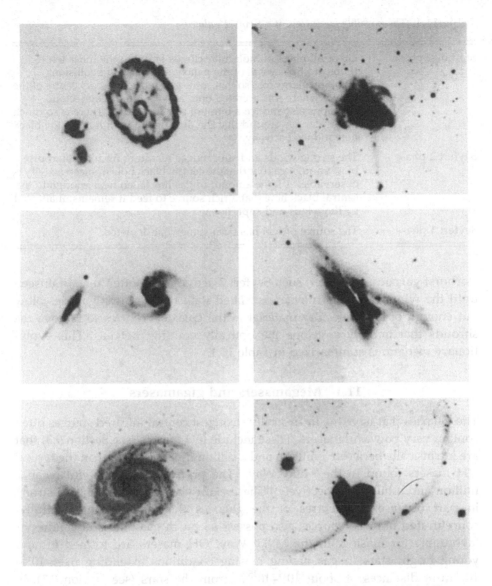

Fig. 11.1. Colliding Galaxies. (From J Barnes, L Hernquist and F Schweizer, *Scientific American*, August (1991).)

Observations of CO emission from Seyfert galaxies show that on average the CO emission from a type 2 Seyfert is about four times as strong as that from a type 1 Seyfert, which typically possesses a CO emissivity comparable to that of an ordinary (i.e. nonactive) spiral galaxy. This result has led to the suggestion that a type 2 Seyfert is powered by the accretion of gas from the reservoir of molecular material detected in CO emission onto the central black hole. Therefore, as the reservoir is depleted the Seyfert 2 galaxy should evolve into a Seyfert 1. A further speculation is that starburst galaxies are the progenitors of Seyfert 2 galaxies and that type 2 Seyfert central activity is occurring in many

Table 11.1. *The possible evolution of a Seyfert galaxy.*

Starburst phase	Great numbers of molecular clouds spiral to the inner ten thousand light years of the galaxy. Cloud–cloud collisions result in energy loss so that some gas gets to the very centre of the galaxy where it is accreted onto a black hole. Cloud–cloud collisions also induce a burst of star formation. There is so much molecular gas around that the dust mixed with it hides the black hole induced activity.
Seyfert 2 phase	The strong winds and supernovae resulting from the starburst blow enough gas out that the central black hole induced activity is observable. A large amount of gas has fallen near enough to the central black hole that a rich source to feed it remains unaffected by the winds and supernovae.
Seyfert 1 phase	The source of fuel has been somewhat depleted.

starburst galaxies. However, such Seyfert 2 activity may often remain unseen until the winds and supernovae associated with the starburst galaxies blow out enough of the molecular material in the galaxies' centres to remove the shrouds that initially envelope the optically emitting hot gas. This evolutionary scenario is summarized in Table 11.1.

11.4   Megamasers and gigamasers

The galaxies that have the intrinsically strongest central infrared sources often contain very powerful masers. These include OH masers (see Section 9.3) that are intrinsically between a million and a billion times brighter than the typical OH masers found in the Milky Way. (The prefixes *mega* and *giga* mean a million and a billion respectively.) These extra-Galactic masers are so strong in part because the centres of the galaxies that have strong centrally-concentrated infrared sources also possess so much gas that is at conditions favourable for masing. In the Milky Way, OH masers are formed around young bright stars in gas having masing conditions extending over 10^{12}–10^{13} m at distances of about 10^{14}–10^{15} m from the stars (see Section 9.3). In the centres of the active galaxies, on the other hand, parcels of gas that extend over distances of roughly one thousand light years (about 10^{19} m) possess properties that produce masing. The total intrinsic brightness of a galactic centre containing amongst the strongest OH masers is roughly 100 million times larger than that of very bright young stars. Perhaps, it isn't too surprising, therefore, that the ratio of the power of the strongest extra-Galactic OH maser to the strongest Milky Way interstellar OH maser is also about 100 million.

Observations of many megamasers, and of the infrared emission from the galaxies in which they are found, have led astronomers to an empirical relationship between the strength of an extra-Galactic OH maser and the total infrared luminosity of its galaxy. This relationship is that the intrinsic strength of the

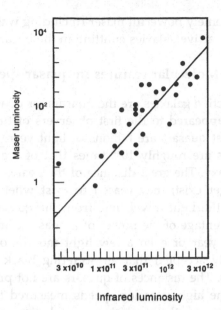

Fig. 11.2. The observed correlation between OH megamaser luminosity and the host galaxy infrared luminosity. The maser luminosity and the infrared luminosity are each measured in units of the total luminosity of the Sun. (From C Henkel, W Baan, and R Mauersberger, *Astronomy and Astrophysics Review*, vol. 3, p.47 (1991).)

maser emission increases as the square of the intrinsic strength of the infrared emission, as shown in Fig. 11.2; i.e. if one galaxy is intrinsically twice as bright as another, then the maser (if any) that it contains will be four times as bright as the maser in the intrinsically fainter galaxy.

This relationship may turn out to be of considerable help in studying the expansion of the Universe. A property of the Big Bang is that all objects, except for those quite near one another (i.e. within the same cluster of galaxies), recede from one another giving rise to recession speeds that are biggest between objects that are farthest from one another. This speed–distance relationship has been studied empirically for objects up to about a few billion light years away. A recession speed is determined from the frequencies by which spectral features are shifted to 'redder' or longer wavelengths by the recession (see Sections 6.5 and 11.3). A distance to an object is established from the ratio of the apparent brightness to the intrinsic brightness. Unfortunately, the intrinsic brightnesses of many of the distant galaxies actually evolve as the galaxies age, so our estimates of distances to the galaxies are quite unreliable. However, a means of setting a reliable distance to a very distant galaxy containing a gigamaser may prove at hand. The ratio of the apparent brightness of the host galactic centre to the apparent gigamaser brightness gives the intrinsic infrared brightness of the source if the empirical relationship shown in Fig. 11.2 is used. The ratio of the apparent and intrinsic infrared brightnesses gives the distance. Thus, gigamasers may be useful in studying the nature of the Universe's expansion.

159

Other types of extremely powerful maser (including water) are found to be associated with many active galaxies emitting in the infrared.

11.5   Molecular features in quasar spectra

The most powerful active galaxies are the quasars, or quasi-stellar objects, so called because they appeared to the first observers of them to be like very blue stars. The nearest quasars are billions of light years from the Sun, and their intrinsic powers are roughly 10^{13} times that of the Sun, or about 100 times that of the Galaxy. The great distance of the nearest quasar means that such objects no longer exist; they ceased to exist when the Universe was younger by roughly the light travel time from the quasar to the Earth. All but a very small percentage of the power of a quasar comes from the region within about a light year or even a few light months of the centre of the quasar. It is generally accepted that an accreting black hole is the power house of such objects. The distances of quasars are not precisely known, but are estimated from the high recession speeds measured from their redshifts and an extrapolation of the recession speed–distance relation known empirically up to several billion light years. Clearly, uncertainties in distances calculated in this way lead to uncertainties in estimates of their intrinsic powers from their apparent brightness.

For about two decades after the thorough study of quasars began, the natures of the galaxies containing them were unknown. For instance, it was thought at one time that quasars were to be found only at the centres of galaxies that were very poor in gas and dust. However, the direct detection of CO emission from the galaxy around a quasar was made in the late 1980s, showing that galaxies that are probably similar to the Milky Way were sometimes hosts of the active centres that created quasars.

Since they are at such great distances from us, quasars often lie behind galaxies and other astronomical objects that are nearer to the Sun. Absorption lines arising in the intervening gas are detected in the quasar spectra. For over two decades no absorption due to molecular hydrogen was seen in these spectra. Yet we know that H_2 is present both when galaxies are forming and when young galaxies are interactive with their surroundings (see Chapter 4). In the late 1980s, the first reliable positive detections of H_2 in absorption against quasars were made. These quasar H_2 absorption line data obviously place some constraints on our understanding of galaxy formation. However, these are not fully understood at the present time. It is clear, nevertheless, that a complete knowledge of molecular astrophysics in all its roles will help us to address many of the challenging problems of astronomy.

Selected references

Black, J H: 'Molecules at early epochs', in *Molecular Astrophysics – A Volume Honouring Alexander Dalgarno*, ed. T W Hartquist, Cambridge University Press, Cambridge (1990).

Selected references

Blitz, L: 'Molecules in galaxies', in *Molecular Astrophysics – A Volume Honouring Alexander Dalgarno*, ed. T W Hartquist, Cambridge University Press, Cambridge (1990).

Henkel, C, Baan, W A and Mauersberger, R: 'Dense molecular gas in galactic nuclei', *Astronomy and Astrophysics Review*, vol. 3, p.47 (1991).

12

Epilogue

—

Optical emission spectroscopy of many types of astronomical source is based on a rather small number of observable lines. Often about a half dozen emission lines of H, N^+, O^{2+}, and S^+ are the only observable features in which optical mapping can be made of a star forming region similar to the Orion Nebula shown in Fig. 1.1, or of a remnant of stellar evolution like the Helix Planetary Nebula shown in Fig. 8.5. In many cases, an additional problem with optical emission spectroscopy is that it does not probe the regions where the most interesting activity occurs because those regions are obscured by dust or are too cool to emit strongly in the optical region. Optical spectroscopy historically provided the foundation of modern astrophysics concerned with the mechanisms and workings of sources rather than simply with their positions and morphologies; however, the advancement of astrophysics has required that techniques complementing traditional optical ones become available.

Radio, millimetre, submillimetre, and infrared spectroscopy have added many of these supplementary techniques. Though most molecules have ultraviolet spectrai features and a small number have detectable optical lines, the lower-frequency, longer wavelength regimes from the infrared to the radio are those in which most of the detected astronomical molecular emissions occur. Since radiation at these frequencies and wavelengths can escape from regions that are highly obscured optically, observations of molecular emissions naturally constitute an important means by which optically obscured sources (whether they are star forming regions, highly evolved stars, or active galactic nuclei) can be studied.

In many sources emissions from 10 to 20 or more separate molecular species can be mapped. The observed compositional variety associated with the chemistry compared to the rather limited data obtainable about the atomic and ionic composition of many objects studied only in optical emissions provides a wealth of information about a thoroughly studied molecular source. In addition, each of many molecular species has a number of accessible emission features; the detection of several lines originating from the same species gives even greater diagnostic opportunities.

The wealth of knowledge derivable from observations of molecules in astronomical sources can lead to an understanding of the basic structures and dynamics of some sources that is deeper and more detailed than that of the structures and dynamics of some sources that can be studied only in optical emission. Molecular emission data may prove to be sufficiently abundant stores of information to have utility in the analysis of basic physical processes operating in many types of astronomical source but that have, until now, been enigmatic. An example of such a process is the mixing across a magnetized turbulent boundary layer between colder gas and hotter, more dynamic, more tenuous gas; an enormous variety of astronomical sources are clumpy and the processes that occur in the boundary layers between clumps and hot gas are of tremendous importance for the mass, momentum, and energy content of the hotter gas, and, thus, for the overall properties of the sources. Another example of such a process is the one that generates the friction in accretion disks, which occur around many binary star systems containing a white dwarf (see Section 8.4), neutron star, or black hole and around the 100 million solar mass black hole at the centre of an active galaxy; molecular emission line data for accretion disks around low mass stars (see Fig. 7.1) will give not only an idea of how our Solar System was born, but also of the basic mechanisms determining the evolution of accretion disks in general. The strongest observed molecular emissions and prominent detectable molecular species often include those that are most influential in the chemical control of the structures and dynamics of the sources in which they originate. Hence, the *diagnostic* studies based on molecular emissions, which we have emphasized so strongly in this chapter, also reveal the *controlling* roles of these chemicals. For instance, HCO^+, the most abundant molecular ion in star forming regions and which is important for the magnetic properties of those regions, has several detectable emission lines which can be mapped.

Until recently, a significant barrier in molecular astrophysics has been the absence of high-angular-resolution maps of molecular emissions. The highest-angular-resolution single dish millimetre line observations have angular resolutions around 30–40 times lower than is typically obtained in routine optical observations. However, as mentioned in Section 1.2, several arrays of dishes designed for the reception of millimetre wave signals are now operational. (The data represented in Fig. 7.1 were obtained with one of these arrays.) The availability of high-resolution maps of molecular emissions will increase in the near future, and molecular astrophysics is at the threshold of an exciting era.

Index